THE TVs *of* TOMORROW

synthesis

A series in the history of chemistry, broadly construed, edited by Carin Berkowitz, Angela N. H. Creager, Ann Johnson, John E. Lesch, Lawrence M. Principe, Alan Rocke, and E. C. Spary, in partnership with the Chemical Heritage Foundation

THE TVs *of* TOMORROW

How RCA's Flat-Screen Dreams Led to the First LCDs

BENJAMIN GROSS

The University of Chicago Press
Chicago and London

The University of Chicago Press, Chicago 60637
The University of Chicago Press, Ltd., London

Published 2018
Printed in the United States of America

27 26 25 24 23 22 21 20 19 18 1 2 3 4 5

ISBN-13: 978-0-226-51997-5 (cloth)
ISBN-13: 978-0-226-54074-0 (e-book)
DOI: 10.7208/chicago/9780226540740.001.0001

Library of Congress Cataloging-in-Publication Data

Names: Gross, Benjamin, author.
Title: The TVs of tomorrow : how RCA's flat-screen dreams led to the first LCDs / Benjamin Gross.
Other titles: Synthesis (University of Chicago Press)
Description: Chicago ; London : The University of Chicago Press, 2018. | Series: Synthesis | Includes bibliographical references and index.
Identifiers: LCCN 2017038282 | ISBN 9780226519975 (cloth : alk. paper) | ISBN 9780226540740 (e-book)
Subjects: LCSH: Television—Receivers and reception.
Classification: LCC TK6653 .G76 2018 | DDC 621.388/87—dc23
LC record available at https://lccn.loc.gov/2017038282

♾ This paper meets the requirements of ANSI/NISO Z39.48-1992 (Permanence of Paper).

In reviewing the past, one's thoughts inevitably turn to the future of electronics and RCA, for it is there that the most interesting history will be written. —Elmer Engstrom

CONTENTS

INTRODUCTION: A WORLD OF SCREENS

On a Tuesday morning in late May 1968, a group of reporters gathered at the New York City headquarters of the Radio Corporation of America (RCA) to await an announcement. Those arriving early perused a press release whose contents could have been extracted from a science-fiction novel. From any other source, the promise of "an all-electronic clock with no moving parts" or "the electronic equivalent of a printed page" might be deemed fantastic, but over the past half century RCA had cultivated an unsurpassed reputation for technological innovation.[1] To many people, the company's establishment of America's first radio network (NBC) and pivotal role in advancing both black-and-white and color television broadcasting confirmed its self-proclaimed status as "The Most Trusted Name in Electronics." Most of those gathered at 30 Rockefeller Plaza that morning were willing to grant RCA the benefit of the doubt even if its latest invention relied on an unfamiliar set of substances with a paradoxical name: liquid crystals.

The reporters' press kits clarified that liquid crystals were relatively common organic compounds with a distinctly counterintuitive combination of characteristics. They possessed the mechanical properties of a liquid—they could be poured and took the shape of their container—but they also retained a somewhat regular molecular arrangement resembling a crystalline solid.[2] Academic scientists had known about liquid crystals for decades before RCA's technical staff took an interest. James Hillier, the vice president in charge of RCA Laboratories and acting master of ceremonies, explained that the company's research into these materials began "several years ago when some of our scientists discovered that an electric field could change the molecular orientation of certain liquid crystals and make them reflect light."[3]

Hillier recalled that although scientists at RCA's main research laboratory in Princeton, New Jersey, found this behavior intriguing, it was slow and only observable within a narrow range of high temperatures. Further investigations soon uncovered another electrical effect, a means of inducing turbulence in a different kind of liquid crystals, causing previously transparent samples to scatter light and take on a milky-white color. The speed and striking appearance of this "dynamic scattering" led to the inauguration of a project to incorporate liquid crystals into electronic displays. The resulting devices were thin, lightweight, and rugged in contrast to the bulky cathode-ray tubes (CRTs) that televisions at that time used to produce images. In addition, since liquid crystals reflected ambient light rather than generating their own, the new displays operated at very low power and would not wash out under bright ambient conditions.[4]

Following technical presentations from members of the Princeton research team, Hillier unveiled several prototype liquid crystal displays (LCDs). These included a light shutter, which could be placed in a door or window "to provide privacy with the push of a button," as well as the promised "all-electronic clock . . . driven by solid state and integrated timing circuits."[5] Perhaps most compelling of all was a high-resolution test pattern, hinting that liquid crystals might one day facilitate the creation of flat-screen, portable televisions (fig. 0.1). "You could take such a set to the beach," Hillier joked, "and, in between bikini watching, see the Mets on TV figure out a new way to lose a ball game."[6]

RCA engineers had confirmed LCDs were physically capable of displaying television pictures, but the engineering hurdles associated

Figure 0.1. George Heilmeier shines a spotlight on a high-resolution liquid crystal display prepared for RCA's 1968 press conference. (David Sarnoff Library Collection, courtesy of Hagley Museum and Library.)

with scaling up their demonstration models to a full-size liquid crystal TV set remained quite high. Hillier warned that it would take time for the company to commercialize the LCD. "We believe we have achieved a true technical breakthrough—a new electro-optic effect—that in the years to come will have a significant effect upon the information handling business," he told the press. "But thus far, it is a research breakthrough. It may be years before even the simplest application we have discussed can be brought to fruition." He hoped that when that occurred, "you will learn about it at an RCA press conference similar to this one."[7]

• • •

Today we live in a world of screens. Electronic displays fill our living rooms, offices, and public spaces, bombarding us with imagery and altering how we consume information and interact with each other. In recent years, no technology has proven as crucial to this transformation as the LCD, which has found a home everywhere from televisions and laptops to cameras and smart phones. Fending off competition from plasma panels and organic light-emitting diodes (OLEDs), the LCD has emerged as the dominant display technology of the digital age and the basis for a multibillion-dollar global enterprise.[8] LCD screens are our primary portals to the data networks that connect us to destinations both real and virtual. They are our on-ramps to the information superhighway. This book took shape on a series of LCDs, and the growing popularity of tablet computers and e-readers makes it increasingly likely that it will be read on one as well.

While the ongoing proliferation of the LCD validates Hillier's predictions regarding its long-term potential, his confidence that his company would stay engaged with the technology proved sadly misplaced. Less than a decade after his presentation, as others embraced the use of liquid crystals in wristwatches and calculator readouts, RCA abandoned efforts to market the LCD. By the time wall-mounted liquid crystal televisions became widely available, the firm itself had succumbed to a 1986 corporate buyout from General Electric (GE). One can still purchase a flat-panel television with an RCA trademark, but it bears only a passing connection to the research organization that introduced the LCD.

The story of liquid crystal research at RCA is one of successful invention and failed innovation. The scientific effort that led to the first LCDs was a triumph, capturing attention from both the popular and scientific press and launching an entirely new sector of the electronics industry. Liquid crystals joined Bell Labs' transistor and DuPont's nylon in confirming the long-standing belief among US corporations that investment in fundamental science would lead directly to the generation of new technologies.[9]

At the same time, RCA's actions showed that faith in this "linear model" of R & D was insufficient to transform its liquid crystal prototypes into products. Before the project's cancellation, the Princeton LCD group fought to retain its independence, receive additional funding, and resolve disagreements with management and manufacturing personnel. Their inability to overcome these obstacles inspired some researchers to adopt an alternative innovation strategy—one that was just emerging when RCA took its liquid crystals public. Following the precedent set by the semiconductor industry, they sought out venture capital and established their own LCD start-up companies. Freed from any commitments to RCA's existing product lines, these spin-offs were among the first to bring liquid crystals to the marketplace. Though they never used the term themselves, such firms nevertheless anticipated the rise of the "disruption" mind-set popular among today's Silicon Valley entrepreneurs.[10]

Whether one endorsed the linear model or a more disruptive approach to innovation, the key issue at stake in the early days of the LCD remained constant—how to ensure an invention's successful transition from the laboratory to the factory. It is a riddle that continues to vex businesses in the twenty-first century. For every Apple or Tesla Motors, there are countless companies who have been unable to bridge that divide. Historian Hyungsub Choi has referred to this process of intrafirm technology transfer as "the perennial predicament of high-tech innovation," and it is the focus of this book.[11] While in later chapters I will discuss the extent to which RCA served as the progenitor for liquid crystal manufacturing operations in Japan, South Korea, and elsewhere, my main objectives are to reconstruct the development of the LCD within the institutional constraints of the company's Princeton laboratories and situate that process alongside the changing strategic goals of RCA's leadership and operating divisions.

What distinguishes this story from previous discussions of intrafirm technology transfer is its sustained focus on the chemists, physicists, electrical engineers, and technicians responsible for flat-panel television research at RCA alongside the managers traditionally emphasized in such accounts.[12] The attention bestowed on the latter dates to the initial establishment of in-house corporate laboratories in the early twentieth century.[13] The unpredictable outcomes of industrial research projects drove business leaders to seek out practical methods to nurture new technical breakthroughs and streamline their development. As the ones responsible for implementing those techniques, personnel in upper and middle management were obvious targets for these analyses and an eager audience for their conclusions. Because of their firsthand experience dealing with these questions, those supervisors were also among the earliest authors of the case studies that filled the pages of newly established management journals after World War II.[14]

The preponderance of books and articles written from the managerial perspective provides one explanation for the relative paucity of historical studies of corporate science written from the viewpoint of industrial researchers. Though many research managers achieved their positions following years in the laboratory, the higher they climbed up the corporate ladder, the more removed they were from the day-to-day realities of the workbench. Complicating matters further is the fact that published depictions of completed projects are more accessible to historians than the laboratory notebooks or internal memorandums that cast innovation as an evolving and highly contingent process. Due to concerns about intellectual property litigation, corporations often keep these documents under lock and key or destroy them altogether. The recent popularity of digital archives has raised additional anxiety over the preservation of research records, both old and new.[15]

Beyond these archival considerations, there is also a practical reason for historians to turn their attention toward industrial managers. The growing number of participants in R & D projects throughout the twentieth century renders it nearly impossible to enumerate each individual's role without devolving into incomprehensibility. Limiting one's scope to a handful of key policy makers supplies a means of imposing structure on an otherwise overwhelming cast of historical actors. Unfortunately, adopting this framework risks reinforcing a long-standing misconception first propounded by the influential sociologist of science, Robert Merton. Merton and his disciples believed that industrial science—with its emphasis on financial profit, rigid organizational structure, and reliance on secrecy to maintain a competitive edge—was qualitatively distinct from research conducted in a university setting. They published numerous studies in the 1950s and 1960s that "started from the presumption that there existed *as a matter of fact* a fundamental conflict in the goals and values of scientists and businesspersons," a view which discouraged critical engagement with corporate laboratories or the managers who ran them.[16]

Recent scholarship has discredited this vision of "pure" science unbound from authority figures and untainted by individual ambitions, but evidence of Merton's impact can be seen in the meager amount of humanities-based scholarship exploring industrial science.[17] According to a National Science Foundation report, in 2013 slightly more than half of American scientists were employed in for-profit settings.[18] The percentage of historical studies examining scientists and engineers working in corporate laboratories, though rising, remains far less.

• • •

The chapters that follow will by no means correct the historiographic imbalance between academic and industrial science. Nor will this book attempt to offer a management-free description of the LCD's origins. Instead it shall strive to present members of RCA's technical staff alongside executives, research managers, and marketing experts in Princeton, New York, and elsewhere as contributors to industrial decision making. Rather than treat science as a component of corporate strategy, I will cast these scientists and engineers as active participants in its formulation, much as Jefferson Cowie revealed organized labor's role in RCA's successive decisions to relocate its television manufacturing facilities.[19] To be clear, this involvement did not imply the absence of a power differential between the laboratory and the boardroom. Even during those periods when they were granted near-total autonomy to set their own research agendas, members of RCA's technical staff could not singlehandedly reallocate money or manpower to their projects or force the operating divisions to establish product lines based on their ideas.

These limitations did not prevent RCA scientists and engineers from exerting control over the innovation process. Both the technical aspects and the overall success (or failure) of the company's flat-panel display projects hinged on their actions. If they felt strongly that a particular line of inquiry merited further attention, they could sidestep the bureaucratic chain of command. Often RCA researchers set aside time between officially sanctioned experiments for pet projects. Especially promising results might lead them to reach out to colleagues, at times crossing organizational lines between research groups or operating divisions to brainstorm next steps, procure materials and equipment, or obtain technical assistance. These alliances, suitably cultivated, enabled scientists to present a stronger case that management should support their work. Should their bosses defer, citing a lack of funding, researchers could also draft contract proposals to secure sponsorship from outside firms, the military, or other government organizations.

From the 1951 speech where RCA chairman David Sarnoff first described what he later termed "picture frame television sets" to the final sale of the company's LCD operation in 1976, RCA's technical staff mobilized all of these tactics in pursuit of a flat-panel successor to the CRT.[20] The short-term gains achieved through such means, however, paled in comparison with their capacity to guide strategic thinking about the subject throughout the corporation. For as much as high-ranking executives like Sarnoff or Hillier cast themselves as authorities when it came to the future of electronics, they depended on the hands-on expertise of company scientists and engineers to evaluate new technologies.

Such insight was particularly valuable when dealing with flat-panel displays, which relied on materials, like liquid crystals, whose physical and electrical characteristics were not fully understood. The investigations that led to the LCD coincided with the postwar boom in industrial chemistry that brought us such products as polypropylene, Kevlar, and Gore-Tex.[21] A similar expansion in materials research was underway at America's electronic firms, one focused less on petrochemicals and more on semiconductors. Organic chemistry, the study of substances containing carbon (often combined with hydrogen, nitrogen, or oxygen), was the province of companies such as DuPont and Dow, while RCA and its peers concentrated on inorganic substances, most notably silicon.[22] Liquid crystals were an exception to this rule, and it took time for RCA chemists and engineers more comfortable with inorganic phosphors to familiarize themselves with the properties of these organic compounds and assess their technical value.

Gradually, those researchers deduced what historians of technology Christophe Lécuyer and David Brock refer to as the "material logic" of liquid crystals: how they behaved, the processes for manipulating them, the constraints their properties imposed on their use, and the creative solutions required to circumvent those difficulties.[23] Through direct consultation, the composition of written reports, or the construction of prototypes, RCA staff members established the boundaries of technological possibility within which the company's managers could maneuver, irrevocably shaping their rhetoric and actions.[24]

Consider, for example, the circumstances surrounding the decision to disclose the existence of LCDs. In the spring of 1968, RCA was in the midst of an identity crisis. David Sarnoff had recently stepped down as CEO, leaving his son Robert in command of the company. Less technically inclined than his father, Robert Sarnoff sought to remake RCA's corporate image and expand its business interests beyond electronics. Realizing that the younger Sarnoff's strategy threatened the company's R & D budget, personnel in Princeton—including James Hillier—wanted to demonstrate the value of maintaining a well-funded central laboratory.

Enter electrical engineer George Heilmeier. It was Heilmeier who, after observing dynamic scattering in 1965, had persuaded his supervisors to support further research into liquid crystal technology. For three years, he had steadily built up the Princeton LCD group, working in secret to avoid alerting any potential competitors. Still, he was eager to reveal what he and his colleagues had accomplished. In monthly reports distributed to senior staff, he supplied detailed summaries of his group's labors and went out of his way to emphasize the wide range of applications that could benefit from liquid crystal readouts. Although the LCD could be seen as merely the latest entry in RCA's ongoing roster of flat-panel display alternatives, Heilmeier's updates persuaded Hillier that it deserved special treatment. Their motivations varied—Hillier recognized that the LCD possessed symbolic value as an exemplar of the type of interdisciplinary collaboration that was only possible at a world-class research center, while Heilmeier wanted to showcase his team's technical ingenuity—but the outcome was the same: a high-profile rollout with top executives in attendance.

Heilmeier's influence over RCA's management was not, however, confined to convincing them to promote liquid crystals. In the weeks before the event, Hillier and RCA's public relations department turned to the technical staff for assistance translating the science behind dynamic scattering for general consumption. To that end they requested that Heilmeier prepare a "writeup . . . describing liquid crystals in semitechnical language—a fact sheet," which would also include a brief history of the project and its participants.[25] The resulting four-page document compressed a complex multiyear process into a streamlined abstract that formed the basis for the earliest stories of the LCD's origins.[26]

When Hillier stood before the crowd of reporters at 30 Rockefeller Plaza and outlined the history of liquid crystal research at RCA, he followed Heilmeier's lead. Neither the official fact sheet nor Hillier's speech divulged the original motivation to examine liquid crystals before knowing about their electro-optic behavior. Furthermore while both Heilmeier and Hillier alluded to the synthesis of room-temperature liquid crystal mixtures as integral to the fabrication of practical displays, neither of their presentations called attention to the other technical issues that had to be resolved before taking the LCD public.[27] The initial observation of dynamic scattering was framed as the self-evident result of a single experiment rather than the product of an extended series of investigations as Heilmeier and his colleagues struggled to reconfigure their liquid crystalline materials for use in displays.

Perhaps the most noteworthy discrepancy between Hillier and Heilmeier's respective rundowns concerned how many researchers

Figure 0.2. The core group of researchers identified in RCA publicity materials as responsible for the development of the LCD. *Left to right:* Lucian Barton, Joseph Castellano, George Heilmeier, Joel Goldmacher, and Louis Zanoni. (Courtesy of Louis Zanoni.)

received credit for developing the LCD. With nearly twenty people involved in the main project in Princeton and several contacts at other RCA operating divisions, some omissions were to be expected. Heilmeier's fact sheet attributed the new displays to a core group of five researchers (fig. 0.2). Hillier's remarks expanded this roster to include a handful of other engineers responsible for assembling the various demonstration devices for the press conference.

In this instance, Heilmeier's version prevailed, as reporters and RCA's in-house publications latched on to his core group, allowing the specific contributions of Hillier's additions—and the others left unmentioned in the ensuing flurry of newspaper and magazine coverage—to fade into the background.[28] The most prominent exception was engineer Robert Lohman, who was photographed comparing his wristwatch to the dynamic scattering clock and who

Figure 0.3. RCA engineer Robert Lohman comparing his mechanical wristwatch and the world's first LCD clock. (David Sarnoff Library Collection, courtesy of Hagley Museum and Library.)

subsequently made appearances in the *New York Times* and *Washington Post* despite being only loosely affiliated with the LCD project (fig. 0.3).[29] Hillier did not go out of his way to correct this narrative. There were other participants, but the people on Heilmeier's list were indisputably vital to the liquid crystal initiative's success and embodied the interdisciplinary spirit of innovation that he sought to illustrate.

Hillier and Heilmeier approached the construction of a popular account of RCA's liquid crystal program with different objectives in mind, but the process in which they jointly engaged remains a common one for scientists, engineers, and their supervisors. Whether employed in a corporate research center or a university lab, there are always decisions to be made when presenting scientific findings. Speeches, journal articles, features in popular periodicals, patents—each genre of communication targets different people and possesses its own set of norms and expectations. Depending on the circumstance, one chooses to highlight certain aspects of the investigative process and downplay others. Where articles frame research as a collaborative endeavor through citations of earlier publications, a patent takes pains to avoid referencing previous work so that its author can retain a claim to originality. In most cases, experimental setbacks or dead ends are not addressed, which allows authors to emphasize the results they deem most relevant but also generates an incomplete chronicle of the discovery process.[30]

These changes and the rationales behind them remain invisible to the audience. Like fellow illusionists witnessing a magic show, other scientists learning about RCA's liquid crystal displays might have recognized that they were privy to just a portion of the story even as the general public took it all at face value. And much like a magic show, the only way to comprehend what really occurred on stage is to go behind the curtain. We must talk to the performers, examine the tools of their trade, and perhaps travel to their workshops to understand how they decided what to include in the act. We must examine the behavior of managers and researchers alike, for both were involved in determining which technologies would leave the laboratory.

• • •

In many ways, RCA's liquid crystal display program and its predecessors are ideal case studies to address these questions. None of the projects in this book approached the scale of Cold War big science, which could involve hundreds of participants. The Princeton LCD group never exceeded two dozen people, a large but not overwhelming cast of characters, most of whom maintained detailed records of their activities. Until its closure at the end of 2009, their notebooks, photographs, and technical reports remained on site at the David Sarnoff Library, which continued to serve as RCA's technical archive after the corporation's sale to GE. These materials, which have since been relocated to the Hagley Museum and Library in Wilmington, Delaware, permit at least a partial disaggregation of their research and interactions with management.

Sadly, the extensive documentary collection at RCA's Princeton facility cannot be matched by the company's other operating divisions. Few written records survive from the liquid crystal assembly lines overseen by RCA's semiconductor division in Somerville, NJ, and still less remains concerning the deliberations of company executives in New York. Fortunately, we have access to the testimony of project personnel, both in Princeton and elsewhere, that can fill in some of these gaps and offer further insight into the shifting status of flat-panel display research within the corporation. Some of these reminiscences were published as books or articles, while others were preserved as oral histories collected under the auspices of the Institute of Electrical and Electronics Engineers (IEEE) History Center, the Smithsonian Institution, the Computer History Museum, the Charles Babbage Institute, and the Chemical Heritage Foundation. Interviews that I conducted while researching this book also opened the door to new, previously unexamined collections of personal papers and artifacts that shed light on efforts to commercialize the LCD, both at RCA and Optel, a spin-off firm whose staff included several members of Heilmeier's team.

Viewed as a whole, these materials allow us to consider the process of industrial innovation from multiple perspectives within RCA over the course of a quarter century. The resulting story is not comprehensive. Even on those rare occasions when the documentary record is complete, historians, like scientists, must make choices about what threads to weave into their narrative tapestries and which to snip away. My hope is that the additional details gained by placing equal importance on corporate researchers and managers will outweigh the omissions. It is by taking this step that we are able to trace the twisting pathways that culminated in the emergence of the LCD to their unlikely source: a luncheon held in a New Jersey cafeteria.

1 THE QUEST FOR MAGNALUX, 1951–1956

It was, by all accounts, a most unusual birthday party. The guest of honor turned sixty in February but had arranged for the celebration to be held in late September. Furthermore, while he spent most of his time in Manhattan, the event took place in the far less urban, though no less civilized, surroundings of Princeton, New Jersey. The festivities themselves were large enough to receive press coverage, yet the participants who garnered the greatest attention did not actually make a physical appearance. Still, absence did not prevent both major candidates from the previous presidential election, Harry Truman and Thomas Dewey, from sending congratulatory telegrams to David Sarnoff, the leader of the Radio Corporation of America (RCA), on an occasion the executive deemed more significant than his birth in a Russian shtetl: the forty-fifth anniversary of his career in electronics.[1]

Much had changed since 1906, when the young Jewish immigrant had taught himself Morse code as an office boy at the American branch of the Marconi Wireless Telegraph Company. That firm had long since vanished, purchased by General Electric (GE) as part of the government-sanctioned business maneuvers that gave birth to RCA.[2] Sarnoff, in turn, had risen through the new company's ranks, using his growing authority to secure RCA's independence from shareholders at GE and Westinghouse.[3] At the same time, he shifted the firm's focus from international and maritime communications to commercial broadcasting and expanded its manufacturing infrastructure through the acquisition of the Victor Talking Machine Company in Camden, a Westinghouse lamp factory in Indianapolis, and a GE vacuum tube plant in Harrison, a suburb of Newark.[4] He also presided over the growth of the company's research facilities, culminating with the September 1942 opening of a laboratory in Princeton, halfway between the factories in Camden and Harrison.[5]

Exactly nine years later, Sarnoff returned to Princeton as the chairman of RCA's board of directors, though he was more proud of the rank of brigadier general, which he had earned for coordinating military and press communications during the D-Day invasion. The company's focus had shifted from radio to television, but even during the Great Depression, Sarnoff's commitment to his technical staff remained strong. Now they wished to repay the favor. Following a celebratory lunch in the laboratory's cafeteria, RCA executive vice president Charles Jolliffe unveiled a plaque officially renaming the Princeton facility the David Sarnoff Research Center (fig. 1.1).[6]

In response to this announcement, Sarnoff, or "the General," as he preferred to be known, delivered a speech that would become the subject of short-term press speculation and long-term reflection among RCA's staff.[7] After thanking his colleagues, he noted that his family was in attendance and that "it is not regarded improper, in the intimacy of one's own family, to make a suggestion occasionally or to throw out a friendly hint about the kind of present you would like for your birthday or your anniversary."[8] With that in mind, Sarnoff proceeded to describe three gifts that he wanted RCA's scientists and engineers to create in time for the 1956 celebration of his golden anniversary in radio.

Sarnoff presented these technologies as "essential inventions for which there is a basic public need" and predicted that they "would expand existing industries and create new ones."[9] Certainly both of these descriptions could apply to one of his birthday gifts, the "Magnalux" light amplifier, which inspired researchers in Princeton to construct RCA's first flat-panel display prototypes. A combination

Figure 1.1. David Sarnoff stands in front of the bronze plaque that officially renamed RCA's Princeton research center in his honor. (David Sarnoff Library Collection, courtesy of Hagley Museum and Library.)

of technological, institutional, and economic factors prevented their inventions from entering commercial production, but these preliminary attempts forced scientists and engineers to consider the practical challenges that had to be resolved to bring the General's dream to life.

Examining their efforts also reveals how members of RCA's technical staff transformed Sarnoff's conception of light amplification. The development of Magnalux was not a simple, unidirectional process in which an executive ordered his workforce to create a specific technology within a discrete time frame. Instead, researchers capitalized on the ambiguity of Sarnoff's request to pursue several projects, each of which embodied different aspects of his vision. Over the next five years, Sarnoff and his fellow managers would alter their speeches and articles about Magnalux to reflect these engineering accomplishments. The public nature of those rhetorical shifts presents a marked contrast with RCA's later flat-panel initiatives, including the liquid crystal display (LCD) program, which were conducted in relative secrecy. As a result, the Magnalux story provides a unique opportunity to witness industrial scientists and engineers visibly shaping discussions of R & D strategy.

THE GENERAL'S REQUEST

While internal accounts of Sarnoff's 1951 remarks accepted his introductory assertion that "I have not prepared a speech for this occasion" at face value, the fact that his comments extended over thirteen single-spaced pages in a subsequently printed commemorative pamphlet suggested the exact opposite.[10] Indeed, as an immigrant, Sarnoff prided himself on his mastery of the English language, both as an external signifier of his assimilation into American society and as a means of exerting authority. "In his view, his speeches were the stuff of history," Sarnoff's biographer Kenneth Bilby observed, "and he crafted them with infinite care, often devoting hours to chiseling a single phrase."[11]

Considering this customary attention to detail, Sarnoff's birthday gift requests come across as vague, particularly in the case of the light amplifier. Unlike the other two presents—a video recording system using magnetic tape (the "Videograph") and an air conditioner without moving parts ("Electronair")—the Magnalux had no obvious precedent. On the surface, the General's wish seemed clear enough:

> One of the presents I would like to have you invent is a true amplifier of light. I have been talking about that for some years and I can get into very animated technical discussions with scientists and engineers as to whether there is such a thing or not. . . . But I think we can all agree that, while we have learned how to amplify electricity, we have not yet learned how to amplify light. . . . Now I should like to have you invent an electronic amplifier of light that will do for television what the amplifier of sound does for radio broadcasting.[12]

Much as audio recorders and mechanical refrigerators were models for his other two gifts, Sarnoff argued, the antecedent for the light amplifier was the loudspeaker, which permitted audio signals to escape from operators' headsets and, in his words, "made radio broadcasting the industry that it is today."[13] In a similar fashion, light amplification "would provide brighter pictures for television which could be projected in the home or the theatre on a screen of any desired size."[14] The Magnalux—literally, a "big light"—would improve on the projection systems RCA previously used for closed-circuit television demonstrations. Those setups, Sarnoff pointed out, "can, of course, enlarge pictures optically, but in the process light is lost and the pictures become dimmer instead of brighter. What is needed is a true amplifier of light itself."[15]

Beyond these generalities, Sarnoff left the technical basis of Magnalux a mystery. George Brown, an electrical engineer who later became an RCA vice president, recalled feeling that the General was drawing a false distinction between audio and video signal amplification.

> Since sound in a radio set comes out of the loudspeaker with sufficient volume because the electrical signals corresponding to the sound are amplified by means of vacuum tubes, I assumed that we already had the problem under control[,] for picture signals were also amplified electronically.[16]

In effect, Sarnoff was asking his engineers to reconsider a problem they had previously solved. But while the vacuum tube and its solid-state analogue, the transistor, could boost the strength of a video signal, neither could increase the brightness of the resultant image as the General had requested.

Given his passion for predicting future developments in communications, it is possible that Sarnoff had no specific technology in mind while drafting his talk. Yet throughout his career, he maintained close relationships with researchers, often traveling to Camden or Princeton for extended visits to discuss the latest electronic advances.[17] Moreover, both George Brown and Elmer Engstrom, the head of RCA Laboratories during the 1950s, asserted that Sarnoff regularly contacted them for technical advice before composing his speeches.[18] Assuming he did so before the 1951 ceremony, his omissions may have been intended to avoid restricting the creativity of RCA scientists or accusations that his ideas were infeasible. More importantly, like any good prophet, Sarnoff recognized the wisdom of embracing a degree of ambiguity to accommodate the unexpected in light of the rapid changes underway within RCA's newly renamed research center and the American electronics industry.

A LABORATORY IN TRANSITION

Sarnoff's birthday gift request came at a tumultuous moment in RCA's history. Retreating from the near-total military orientation of its research facilities during World War II, by 1951 RCA found its hopes to dominate the postwar consumer electronics market threatened by the Columbia Broadcasting System (CBS).[19] Sarnoff and his opposite number at CBS, William Paley, had been engaged in a struggle to establish technical standards for American television broadcasts since before Pearl Harbor. RCA's subsidiary, the National Broadcasting Company (NBC) commenced regular TV service in 1939 with the intent of using the profits from monochrome sets to develop a color system. In 1940, Paley undercut these plans by announcing that CBS had devised a market-ready color television apparatus, which projected images through a rotating set of color filters. Unlike RCA's planned "compatible color" system, whose electronically generated images could still be viewed, albeit in monochrome, on existing black-and-white sets, the CBS system would render the nearly twelve million televisions sold between 1946 and 1950 obsolete.[20]

With the future of their industry at stake, Sarnoff and Paley lobbied the Federal Communications Commission (FCC) to establish broadcasting standards favoring each company's respective technology.[21] The FCC postponed discussion of the matter until after the war but in 1950 authorized broadcasts using the CBS system. Sarnoff appealed the ruling, calling the FCC's actions "scientifically unsound

Figure 1.2. Aerial photograph of the David Sarnoff Research Center taken in 1952. (David Sarnoff Library Collection, courtesy of Hagley Museum and Library.)

and against the public interest," but the Supreme Court supported the commission's decision.[22] Fortunately for RCA, the military's requisition of electronics materiel following the outbreak of hostilities in Korea prevented CBS from expanding color set production. The Korean conflict also supplied RCA with an opportunity to refine the compatible color system and confirm its superiority to Paley's "whirling disc."[23]

Consequently, color television retained its status as the highest priority project at the David Sarnoff Research Center (DSRC), a constant objective for a company trying to adapt to the ever-changing landscape of the US electronics market (fig. 1.2). Between 1942 and 1945, the complement of technical personnel in Princeton had remained fixed, as RCA devoted itself to wartime radar and television projects.[24] Now a growing slate of competitors and the increased importance of semiconductor technologies provided incentives to broaden its research base. Sarnoff delegated this responsibility to Elmer Engstrom, who had supervised RCA television research in the 1930s and was now vice president in charge of RCA's laboratories.[25] Engstrom introduced a new training program that exposed incoming staff members to work in different labs before they received their first official assignment. He also worked with Douglas Ewing, a former physics professor who previously managed the development of aerial navigation systems in Camden, to oversee the DSRC's recruitment drive and organize partnerships with nearby universities such as Rutgers and Princeton.[26]

Engstrom and Ewing's actions were the vanguard of a realignment of RCA's R & D strategy, institutionalizing the notion that fundamental research, conducted without any practical outcome in mind, would result in discoveries that might serve as the basis for new technologies. That new knowledge would then leave the laboratory for subsequent applications development at the company's operating divisions in Camden, Harrison, or elsewhere. This assumed one-way progression from basic to applied science, rooted in the writings of electrical engineer turned science policy expert Vannevar Bush, came to be known as the linear model of innovation and became the touchstone for industrial research policy at RCA—as well as firms such as American Telephone and Telegraph (AT&T), DuPont, and International Business Machines (IBM)—through much of the Cold War.[27]

As early as 1945, Engstrom had worried that RCA's wartime research was nearing the limits of scientific understanding. "Technological advances have been rapid," he told his superiors, "and have now reached the point where further progress is being materially held up because of lack of knowledge of basic phenomena."[28] Hiring more researchers, especially those inclined toward theoretical work on up-and-coming subjects such as solid-state physics, which studied the behavior of electrons within materials rather than in a vacuum, was the obvious solution to this deficiency. The company heeded Engstrom's advice. By 1951, the DSRC's annual report boasted that RCA's technical staff had more than doubled in size since the end of World

War II, and its solid-state research program was "probably the broadest in any industrial research laboratory in the country."[29]

The target audience of David Sarnoff's birthday gift speech was therefore one in transition, moving away from the familiar world of vacuum-tube technologies and into the enigmatic realm of the solid state. Attempts to develop the General's light amplifier followed a similar trajectory. Engstrom and Ewing's emphasis on fundamental research and the corresponding marginalization of product development in Princeton would also affect the DSRC's organizational standing within RCA and the eventual fates of the Magnalux prototypes developed between 1951 and 1956.

EMBRACING ELECTROLUMINESCENCE

Despite Sarnoff's attempts to distance Magnalux from existing projection setups, RCA engineers' familiarity with that equipment rendered them a logical starting point in the search for a practical light amplifier. Those systems utilized cathode-ray tubes (CRTs), the functional core of RCA's conventional television sets. A CRT was a vacuum tube containing a negatively charged electrode (the cathode) that emitted a stream of electrons when heated. In standard televisions, a set of magnetic coils guided this beam of "cathode rays" across a phosphor-coated faceplate to generate a picture. The basic principle of image creation remained the same for projection systems, with the light from the resulting picture passing through a series of lenses and on to a movie screen.[30] Through the use of multiple CRTs, one could superimpose the primary (red, green, and blue) components of an image, creating a picture that one reporter claimed "might easily have been mistaken by a tolerant eye for a fairly well photographed Kodachrome or Technicolor film."[31]

Theater television demonstrations captivated the press during the 1940s and later served as a powerful propaganda tool during postwar debates over color broadcasting standards. These successes suggested the possibility of a home projection system. No one gave the question more serious consideration than David Epstein, one of the original designers of the optics used in RCA's projectors.[32] Epstein conceded that several issues would need to be addressed to make home projection TV possible. The most significant of these, as Sarnoff had indicated, would be counteracting the reduced brightness associated with image enlargement. During public screenings, RCA engineers had solved this problem through the use of high voltages, several times greater than those used in a typical home television, to maximize their system's light output. Further complicating matters, in small spaces any misalignment of the three CRTs required to generate color pictures produced severe optical distortions.[33]

Epstein dealt with the alignment problem using a compact optical system that arranged the three CRTs in a T shape and used special mirrors to selectively filter certain colors of light. Though the compact spacing of these mirrors made the internal optics more complicated, he was nonetheless able to construct a self-contained projection television by 1954.[34] The result more closely resembled a traditional TV console than Sarnoff's Magnalux. Rather than displaying an image on a screen of any size, Epstein's setup could produce pictures no larger than 18 × 24 inches. While its picture resolution and brightness compared favorably with existing televisions, the contrast was dramatically inferior, leading to the project's termination in 1955.[35]

Before then, however, two members of Epstein's research group had started exploring alternative avenues toward light amplification. Frederick Nicoll was a Canadian-born physicist who had helped develop the reflection-free glass used in Epstein's projectors and RCA's wartime radar displays.[36] After transferring to Princeton from RCA's factory in Camden, he joined the compatible color project and soon met Benjamin Kazan, a CRT specialist who came to RCA after a stint at the Signal Corps Engineering Laboratories at Camp Evans, New Jersey.[37]

In 1952, Nicoll and Kazan earned a joint commendation from RCA's management for designing an improved color television picture tube.[38] Their search for a follow-up project eventually led them to light amplification, presumably after learning of a potential solution under investigation at the Massachusetts laboratory of one of RCA's competitors. The company in question, Sylvania Electric Products, was a leading manufacturer of vacuum tubes and other electrical components. Like RCA, Sylvania invested heavily in solid-state research after World War II, but unlike at the DSRC, their researchers decided to explore an obscure effect known as electroluminescence.

Parisian chemist Georges Destriau had first described this phenomenon during the 1930s. While experimenting with different fluorescent compounds, Destriau encountered a group of materials that emitted light when subjected to an electric field. Once he confirmed that his observations had not resulted from accidental exposure to ultraviolet or infrared radiation, he published his findings. Originally, Destriau referred to this behavior as "electrophotoluminescence," but his colleague Maurice Curie (the nephew of Pierre) shortened the name, since the presence of light was not a prerequisite for the effect.[39]

Destriau's experiments elicited mixed responses in the United States. Humboldt Leverenz, the head of RCA's phosphor group, argued in a 1950 textbook that electroluminescence was not a new or distinct effect but rather the result of "electric discharges in gases."[40] His counterparts at Sylvania were more willing to keep an open mind about the Frenchman's results. By 1951, they had synthesized their own electroluminescent phosphors and manufactured glowing green panels suitable for use in night lights, alarm clocks, or other household products.[41] News of these achievements forced Leverenz to abandon his skepticism, and he authorized RCA's chemists to follow suit with an eye toward display applications.[42]

Nicoll and Kazan were among the first to take advantage of the DSRC's new supply of electroluminescent materials. They started by reproducing experiments conducted at Sylvania, eventually creating a panel whose luminance exceeded that of a home television. "The life of a sample at this level is very short with present materials and techniques," they reported, "but the experiment indicates the future possibilities of electroluminescence as a light source."[43] Having confirmed these compounds were bright enough for displays, they now faced the challenge of moving from a single light-emitting element to a device capable of showing an image.

It is important to emphasize that up to this point, Nicoll and Kazan's experiments revolved around image production, not light amplification. Given their previous research interests, these early electroluminescence investigations need not have stemmed from Sarnoff's birthday gift request. Whether the same was true of their next experiment is less clear. What is known is that the two men became aware of ongoing research at the DSRC into another solid-state effect: photoconductivity. Photoconductive substances convey an electric current when exposed to light, a property that RCA scientists had harnessed to create more sensitive video cameras.[44] Nicoll and Kazan believed this same behavior could replace the manual switching used to trigger their earlier samples.

To test their hypothesis, they obtained a small, photoconductive crystal of cadmium sulfide, connected it in series with one of their electroluminescent panels, and applied a voltage to the circuit. When the photoconductor was dark, so was the panel. But as they increased the illumination on the crystal, its resistance dropped, allowing the voltage across the panel to increase and the phosphors to glow a steadily brighter green. Additional tests revealed that depending on the substances involved, the amount of light emitted by the electroluminescent materials could be several times greater than the incident light focused on the photoconductor. As they wrote in 1952, "under steady state conditions, it is possible to obtain some light amplification."[45]

Regardless of whether or not Sarnoff's speech inspired their work, Nicoll and Kazan knew they had discovered a viable technological basis for a solid-state light amplifier, and they accelerated their research over the next year with the goal of producing panels that could generate TV-quality images. Using new photoconductive compounds, by 1953 they constructed a 6 × 6 inch panel that accurately reproduced photographic slides or television images projected on its rear surface in shades of yellow green. The resulting pictures were not perceptibly brighter, and "the slowness of response of the available (PC) [photoconductive] powders . . . caused some smearing of moving objects," but the two men believed that they could resolve these issues.[46]

Kazan and Nicoll spent the remainder of 1953 testing new phosphors and panel structures, ultimately settling on a design that utilized V-shaped grooves to maximize the photoconductor's exposure to light. By 1954, they were able to demonstrate 12 × 12 inch panels that were a quarter-inch thick and able to duplicate incident images with a 15–20-times increase in brightness (fig. 1.3). They could also convert X-rays into visible light, suggesting a possible application in fluoroscopy.[47]

The one shortcoming of the original panel that they were unable to overcome was the smearing of moving images. But, they concluded, as new photoconductive materials became available, "the development of light amplifiers having higher gain and light output than the existing types and with response times sufficient for T-V purposes thus appears as a reasonable possibility."[48]

MAGNETIC LOGIC, DISPLAY DREAMS

On the surface, David Sarnoff had envisioned the refinement of an existing technology—projection television—so that it would be accessible to the home consumer. In contrast, internal documents show that RCA's technical staff interpreted his remarks not as a call for enhanced projectors but for a successor to the CRT itself. The DSRC's annual research report for 1953, for example, prefaced a discussion of Nicoll and Kazan's research with the following observation:

> How long it will be before the cathode-ray tube is replaced by some more effective means for reproducing a television picture is not predictable at this time. However, there is little doubt that this will come about. A program has been under way in RCA Laboratories for some time now to investigate means other than cathode-ray tubes for reproducing color television images. Methods using electroluminescence give indications that they may be particularly attractive especially for the formation of large pictures.[49]

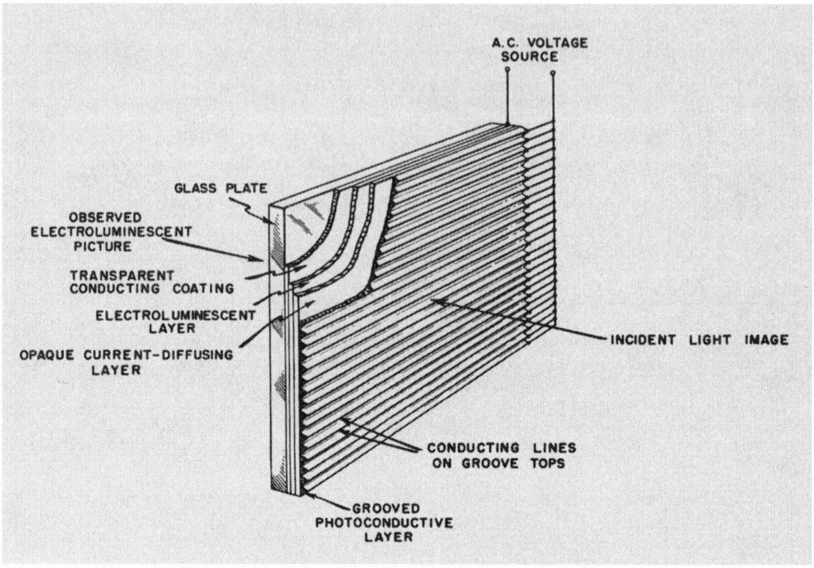

Figure 1.3. Light amplifying panel developed by Frederick Nicoll and Benjamin Kazan, shown in schematic view and in operation. An image projected on the grooved photoconductive layer was reproduced as a brighter picture on the electroluminescent screen. (RCA Laboratories, *Research Report 1954*, 81; RCA publicity photograph. David Sarnoff Library Collection, courtesy of Hagley Museum and Library.)

The roots of the disparity between the General's public statements and the DSRC's technical agenda can be traced to the former's beliefs regarding the specific role of scientific research in the consumer electronics industry. Sarnoff had long embraced obsolescence as an unavoidable aspect of that business. Radio had given way to television, and if he had his way the compatible color system would soon supplant black-and-white sets. The only constant in a technology-driven enterprise, he argued at the DSRC's dedication ceremony, was change, and investment in long-term research projects was the most effective response to that dynamic. "In the RCA, we do not fear or resist change," he explained. "Instead of a wicked ghost that threatens extinction, we see a beneficent wraith whose proddings stimulate opportunity, advance prosperity, and raise the standards of living."[50]

Given this position, it is not surprising that Sarnoff had already started pondering his company's post–color TV strategy. During these deliberations, he probably reached out to research managers such as Engstrom, who would have informed him of promising areas of solid-state research that could facilitate the construction of flat-panel displays. The relative novelty of this work and the risks of inviting additional competition offer plausible explanations for avoiding mention of electroluminescence directly in his birthday gift speech. Instead, he contented himself with oblique allusions to "electrons in solids" and promises of exciting changes on the technological horizon. "My friends, the wireless I knew 45 years ago is not the radio of today," he observed. "The television you know now as pioneers will not be the television of tomorrow."[51]

If it was, in fact, the General's intent to inspire researchers to create "the television of tomorrow," Nicoll and Kazan's light-amplifying panels would not fit the bill because of their inability to present moving images. There were, however, others at the DSRC excited by the prospect of a flat-panel electroluminescent display. Foremost among them was an electrical engineer named Jan Rajchman.

As a wartime consultant for the ENIAC (Electronic Numerical Integrator and Computer) project at the University of Pennsylvania, Rajchman had established himself as an expert in computer memories.[52] Digital computing was a low-priority research topic at RCA, but Rajchman continued to develop new data storage devices, including some of the earliest magnetic core memories. A core memory system consisted of a wire matrix with doughnut-shaped cores made of ferrite, a magnetic ceramic, at each intersection. Through the application of current to a given set of horizontal and vertical wires, one could switch the direction of a chosen core's magnetic field, corresponding to a 1 or 0 in binary code.[53]

Rajchman's interest in television resulted from the realization that the mechanism used to select a specific core in his magnetic arrays could also activate a pixel in a display. Theoretically, one could mimic an electron beam sweeping across a screen by sequentially activating each of the cores in a given row, thereby eliminating the need for a vacuum tube. "For a long time," Rajchman elaborated in a 1952 patent disclosure, "there has been the desire to produce a 'Mural Television Display' which would be a large flat box-like device with a large rectangular face and relatively small thickness. This could then be used on the wall of the viewing room."[54] Such a device could be built using existing magnetic cores and electroluminescent phosphors, but Rajchman knew that more sophisticated equipment would be needed to move from on-off (i.e., black or white) switching to the production of halftone images.

Rajchman sketched out several possible mural television designs before turning his attention to a pair of large-scale computer memory projects.[55] The myriabit (10,000-core) and megabit (1,000,000-core) arrays were intended to force DSRC engineers to create more efficient means of fabricating and wiring an increasing number of magnetic cores, but they ended up supplying Rajchman with the necessary tools to assemble a flat-panel display.[56] The key was his decision to streamline the manufacturing process by switching from individual cores to ferrite plates with evenly spaced rows of holes. As with previous core memories, the magnetic field around each hole could be used to store a bit of information, but these apertured plates were more compact and simpler to produce.[57]

The shift from cores to plates did not occur without incident. "The idea was that an integrated approach was better than an individual approach," Rajchman told interviewers in 1975. "So, for example, we made plates with holes and asked 'How close can you put the holes?' We started to put the holes very close together. When we put the holes very close together, we found that they started to interact."[58] This crosstalk could wreak havoc on computer function, so Rajchman and fellow engineer Arthur Lo launched an investigation to characterize, and thereby minimize, the effect.

One result of these experiments was the 1954 creation of a new magnetic storage element called the transfluxor. In its most basic form, the transfluxor consisted of a ferrite core with two holes of unequal diameter that could mimic the binary switching in Rajchman's computer memories but could also store intermediate values between 0 and 1.[59] In conjunction with an electroluminescent phosphor, it allowed for the production of the variable brightness required for halftone images. With its simple construction and a reliability greater than contemporary transistors, Rajchman and Lo predicted that the transfluxor would "become a basic element in electronic circuits of the future," enjoying widespread use in digital computing, automation, and communication systems.[60]

Curiously, Rajchman omitted any mention of flat-panel displays while extolling the virtues of the transfluxor, but he filled his laboratory notebook with entries outlining an ambitious plan to build a 1,200-element electroluminescent television.[61] He also recruited additional staff, including experimental physicist George Briggs, to assemble a prototype. Together Rajchman, Briggs, and Lo spent the early months of 1955 working on the mural television project, often staying at the labs well into the evening. Each of the 1,200 transfluxors used in the display had to be hand wired and then connected to the addressing circuitry, which they installed on two nearby carts in order to maintain a thin form factor.[62] It was difficult, time-consuming work, made all the more dangerous by the high voltages involved, as Briggs learned one day when he received a nasty electric shock. "I was knocked clean across the room," he remembered. "I couldn't move for half a minute anyhow, and I don't think I was breathing either."[63]

Figure 1.4. Each of the electroluminescent cells in Jan Rajchman's mural television *(above)* was connected to a transfluxor—a multiaperture magnetic core that permitted on-off and halftone switching. The display's resolution was low but still sufficient to reproduce Rajchman's face *(facing page)*. (Jan Rajchman, *Mural Television*, PEM-1013C, 26 Oct. 1955, pp. 8–9. David Sarnoff Library Collection, courtesy of Hagley Museum and Library.)

Whatever injuries the group sustained had healed by April 1955 when Rajchman invited half a dozen people, including Kazan and Nicoll, to his laboratory for a demonstration.[64] The completed mural television display—consisting of thirty rows of forty electroluminescent cells—measured 14 × 18 inches and was only 1.5 inches thick. A specially configured camera generated video signals that appeared on screen as pale green moving images (fig. 1.4). "The picture was uniform, showed good half-tone range and high contrast," Rajchman

reported.[65] The frame rate was only 15 frames/second, as opposed to the 30 used in standard TV broadcasts, but this was sufficient to provide an "adequate illusion of continuous action."[66] The display's transfluxors also allowed users to freeze and store a given image on screen indefinitely without requiring additional standby power.

These technical achievements came at a steep price. With only 1,200 cells, the display's resolution was significantly lower than a conventional CRT. "So much artifice is required at every picture element," Rajchman explained, "that the fabrication of a sufficient number of elements for standard television would be extremely costly and impractical. Also the power consumption and associated electronic circuits are very onerous."[67] Even if heretofore unknown techniques permitted the construction of larger displays, the brightness of existing electroluminescent materials left much to be desired. "You had to have a darkened room to see the meatball [i.e., RCA's circular logo] or

Rajchman's face," Briggs recalled.[68] Theoretically the application of higher voltages could surmount this issue, but as Rajchman's team learned firsthand, doing so led the phosphors to burn out, emitting foul-smelling smoke in the process.[69]

In spite of these flaws, Rajchman's display confirmed that one could produce a moving electronic image without using a cathode-ray tube. The pictures may have been faint and grainy, but Rajchman imagined that devices similar to his prototype might be utilized "in cases where a relatively small number of picture elements is tolerable. . . . Radar displays, outputs of computing systems, [and] displays of numerical and alphabetical information are examples of such applications."[70] Meanwhile, he felt RCA should devote its resources to developing better magnetic switching circuitry and less time-intensive display assembly procedures. "It is believed that the solutions of these essential problems will not only yield a second mural television working model greatly improved with respect to the first," Rajchman concluded," but will also pave the way for still better systems."[71]

TELEVISIONS WITHOUT TUBES?

Nicoll and Kazan's presence at Rajchman's mural television demonstration serves as a reminder that these three men did not conduct their work in isolation. Nor did their shared enthusiasm for electroluminescent television remain confined to the DSRC. Take, for example, Jan Rajchman's May 1955 patent application for an "electrical display device" based on his transfluxor-driven model.[72] This document contained a brief description of "how a mural image reproducer might be utilized in the living room of a modern home":

> It is noted that the mural image reproducer is a flat large image display device which is hung at an appropriate position on a wall, in much the same manner as a favorite painting. The mural image reproducer is flat. Its control cable is coupled in turn to a remotely located control box which would

be utilized to adjust the television channel received and the contrast and brightness of the reproduced image.[73]

The key features of that device corresponded closely to Rajchman's mural television with one exception: the wired control box, which could adjust the channel or other display parameters. Rajchman foresaw the eventual creation of a tabletop controller "small enough to fit into a cigar box," but such a setup was a far cry from the carts of tubes needed to operate his display.[74] The description in his patent was, however, reminiscent of depictions of mural television circulating in the popular press thanks to RCA's leadership.

In August 1954, months before Rajchman's team completed their transfluxor-driven prototype, David Sarnoff delivered a speech in Chicago announcing that "at some time in the future—I will hazard a guess and say five years from now—no tubes will be needed in a television set—not even the picture tube."[75] He restated that prediction, though without an estimated time frame, two months later at a St. Louis celebration of the seventy-fifth anniversary of Edison's lightbulb.[76] In each case, he noted that electroluminescence would allow for larger displays without bulky cabinets. These screens would be "connected directly by a small cable, with a little television box—about the size of an average cigar box that can be placed anywhere in the room."[77] Sarnoff's proposed control mechanism relied on transistors—not tubes!—to manipulate the size and brightness of an image and switch between black-and-white and color.

The General was not alone in advancing such claims. At a 1955 electronics conference in San Francisco, DSRC head Elmer Engstrom asserted that electroluminescent materials and miniaturized electronic circuitry "will give us mural television. Its form will be that of a thin screen decorating a wall and controlled remotely from a small box beside the viewer elsewhere in the room."[78] An illustration of a "picture frame television," published in the July 1955 issue of *Radio Age*, RCA's in-house magazine, also bore a striking resemblance to the invention described in Rajchman's patent (fig. 1.5).[79]

Figure 1.5. Drawing from RCA's *Radio Age* magazine, seemingly inspired by Rajchman's research, suggesting that picture frame televisions would be commonplace by 1975. Another of the General's birthday gifts—a noiseless electronic refrigerator—also makes an appearance. ("Products of the Future . . . ," *Radio Age* 14, no. 3 [July 1955], 16. David Sarnoff Library Collection, courtesy of Hagley Museum and Library. Scan courtesy of Linda Hall Library of Science, Engineering & Technology.)

With the benefit of hindsight, the parallels between Sarnoff and Engstrom's speeches and Rajchman's patent are obvious, but before 1956, neither Sarnoff nor any other RCA executive referred directly to the computer researcher's electroluminescence investigations. The only flat-panel developments brought to the public's attention were Kazan and Nicoll's light amplifiers, which a December 1954 press announcement presented as an interim step toward the creation of "a thin, flat picture screen on the wall."[80] Sarnoff admitted that this new technology was not ready for the market. "You may wonder what philosophy prompts me to reveal these new developments publicly while they are still in the experimental stage," he mused in a January 1955 address before the American Institute of Electrical Engineers. The General's answer was simple: publicizing research was essential to maintaining RCA's technical leadership. "We welcome competition. It spurs our own activities and increases the possibilities for earlier achievement of desired results."[81]

As with his 1951 speech, Sarnoff's straightforward justification concealed a more complex backstory. His paeans to the competitive benefits of promoting cutting-edge research masked how much the rivalries he cited as essential to RCA's long-term success could also provoke actions it might otherwise wish to avoid. For although RCA executives had made passing references to electroluminescence research, Sarnoff chose only to disclose a specific outcome of that work after General Electric announced plans to demonstrate its own light-amplifying panels before the end of 1954.[82]

Nicoll and Kazan's prototypes were not market ready, but RCA's publicity department could not let GE co-opt one of the General's birthday presents. While GE forced Sarnoff's hand, his response demonstrated a keen grasp of strategic information disclosure. In one move, RCA simultaneously regained the initiative from GE, confirmed that Sarnoff's request had produced tangible results, and inspired speculation as to whether RCA's demonstration presaged the unveiling of true "picture-on-the-wall television" at his golden anniversary dinner.[83]

MAGNALUX REVEALED

The ceremonies commemorating Sarnoff's fiftieth year in the electronics industry dwarfed those of 1951 in every respect. More than a thousand guests crowded into the Waldorf-Astoria Hotel in late September 1956 for a formal banquet in the General's honor. Once again the president of the United States and the governor of New York sent celebratory letters, but now Dwight Eisenhower and Averell Harriman were joined by Richard Nixon, Winston Churchill, and Adlai

Stevenson in extending their well-wishes to the man of the hour.[84] RCA itself had grown over the preceding half decade, with profits breaking all previous records. "My completion of fifty years in harness happily coincides with the first 'billion dollar year' in the history of our company," Sarnoff announced. "What a magnificent anniversary present!"[85]

Of course, the members of RCA's technical staff had spent five years preparing several presents of their own. Speaking on behalf of the DSRC, Elmer Engstrom recounted Sarnoff's requests. Now, he proclaimed, "the five years have passed, the day has come, and this is truly a day of fulfillment because we have met your desires and I think in some instances exceeded them."[86] Since not everyone would be able to attend the demonstrations scheduled for the next day in Princeton, Engstrom brought slides of all three gifts in action. As the lights dimmed, he proceeded to describe each device, beginning with the Magnalux.

From newspaper coverage and RCA press releases, a well-read observer sitting in the darkened Waldorf-Astoria ballroom could have deduced Sarnoff's light amplifier would be based on electroluminescence. Those with access to the company's internal research reports would know about Nicoll and Kazan's panels and Rajchman's mural television, but until Engstrom's speech, it would be very difficult to ascertain which of those prototypes had been anointed the embodiment of Magnalux. The former could increase the brightness of an image in accordance with the General's original speech, but the latter aligned more closely with his subsequent predictions about television's future. In an article published in the *New York Times Magazine* the day of his golden anniversary party, Sarnoff had insisted that full-color "picture frame" television sets resembling Rajchman's would soon be within reach.[87] His remarks that evening also declared that electroluminescence would "give us brighter and bigger TV pictures, and ultimately replace the TV tube altogether with a thin, flat-surface screen that will be hung like a picture on the wall."[88]

Yet as soon as Engstrom started his presentation, it became clear that RCA's leaders did not consider Rajchman's mural television ready

Figure 1.6. Benjamin Kazan projects David Sarnoff's portrait on an improved light amplifying panel, which the DSRC presented to the General as part of the 1956 festivities celebrating his fifty-year career in electronics. (David Sarnoff Library Collection, courtesy of Hagley Museum and Library.)

for public scrutiny. Engstrom's first slide featured Kazan projecting a portrait of David Sarnoff on to a "picture frame type panel," which could amplify the brightness of incident light by up to one thousand times (fig. 1.6). The improved brightness of this latest model's phosphors was not matched by a corresponding decrease in photoconductor response times, but Engstrom elided its inability to present TV images by highlighting its use in medical equipment. These panels, he suggested, might allow radiologists to lower the dose of X-rays necessary to diagnose a patient while providing brighter images than existing fluoroscopes.[89]

Once Engstrom completed his presentation, Sarnoff thanked RCA's researchers "for their pioneering courage, their perseverance, their competence unmatched in this field."[90] He acknowledged that the difficulties they had faced while creating a light amplifier—as well as his two other presents—were immense. "But the fact remains that in five short years they have succeeded in turning what were bold dreams and hopes into proud realities."[91] Their successes validated Sarnoff's trust in the linear model and the intrinsic manageability of corporate innovation. This philosophical outlook perhaps explains why Sarnoff retained references to the future promise of electroluminescent television in his speech even though the DSRC's managers did not think that Jan Rajchman's prototype was ready for public demonstrations or commercial production.

THE FATE OF MAGNALUX

The following day, Sarnoff and his entourage traveled to Princeton for the formal presentation of the General's anniversary gifts. Once again, Elmer Engstrom served as master of ceremonies. He first led a delegation of visitors to a room cooled by Electronair, wall-mounted thermoelectric panels that, as per Sarnoff's request, operated with no moving parts. Then it was off to a faux living room setup, where they watched NBC programming recorded using the new Videograph magnetic tape system. Finally, they witnessed a live demonstration of the now familiar Magnalux light amplifier.

To Sarnoff, these presentations showcased "modern science at its best, concentrating its formidable talents upon the constructive task of providing a wealth of devices and techniques for man's well-being."[92] For all his confidence, however, he still hesitated when asked when these gifts would be available to the public. "How long this will take nobody knows," the General told reporters. "As an uneducated guess I'd say that between now and the end of five years you could expect to see practical products embodying these principles on the market."[93]

RCA's researchers took Sarnoff at his word and, after his departure, moved to ensure the prompt commercialization of these technologies. Though Electronair panels proved inefficient for home use, cooling equipment based on the same concept found a place on nuclear submarines, where silent operation was essential and excess power was readily available. The Videograph group reached a cross-licensing agreement with Ampex, a California-based electronics firm that had independently developed its own black-and-white video recorder, to produce a magnetic tape system capable of recording RCA's color broadcasts.[94]

As it had for the preceding five years, the Magnalux project continued along two fronts. Nicoll and Kazan demonstrated their panels at a meeting of the American Roentgen Ray Society in Washington, DC, and worked to standardize their production.[95] Meanwhile, Rajchman and his team, whose transfluxor-driven display never received direct attention from the press, continued to improve their designs. The DSRC's annual report for 1956 described Rajchman's mural television as one of RCA's "major continuing projects" and promised ongoing support "with an increased staff and on a broader front."[96]

Once again, Sarnoff's "uneducated guess" was poised to come true, but before the anticipated five years had elapsed, both electroluminescent display projects were discontinued. RCA personnel and business historians have attributed this outcome to a lapse in the General's technical judgment, which led him to overestimate the performance of electroluminescent displays and the ease with which they might be commercialized.[97] Engineering obstacles definitely contributed to the cancellation of electroluminescence research at the DSRC, but that decision also hinged on a combination of organizational and economic shifts that radically altered RCA's research infrastructure during the second half of the 1950s.

In 1951, Sarnoff had described the DSRC as an institution "which does basic research and applied research, and translates the images of the mind into useful products and services for the public."[98] This holistic conception of R & D may have been accurate in the immediate aftermath of World War II, when close working relationships existed between the corporation's laboratories and operating divisions.[99] But as Engstrom and Ewing steered the DSRC toward more theoretical investigations, those links began to weaken. Once the staff in Princeton had consisted primarily of engineers drawn from RCA factories in Camden and Harrison, but now its ranks swelled with university-trained researchers unfamiliar with the logistics of technology transfer.

The growing schism between RCA's laboratories and factories widened further following Ewing's 1954 appointment as the DSRC's administrative director. Citing a need for greater operational flexibility, Ewing implemented a new "building-block" strategy, transferring nearly all product-related research to the divisions "to allow the research laboratories freedom to work on the more speculative and long-range projects."[100] He also discouraged project-based research teams, preferring instead to form "temporary task groups" drawn on an ad hoc basis from a half dozen new laboratories organized around general concepts such as physical and chemical research, radio research, or television systems research.[101]

The adoption of the building-block strategy cemented the DSRC's reputation within the corporation as a "country club," where funds were squandered on outlandish projects with limited commercial value, like the General's birthday gifts.[102] To personnel working on production lines in Camden or Indianapolis, the money flowing into Princeton was a diversion from the company's chief priority: color television. RCA had finally persuaded the FCC to endorse its compatible color broadcasting standard at the end of 1953. Now that their long battle against CBS was over, the company's manufacturing divisions were eager to recoup their $50 million investment in color TV.[103]

As assembly lines whirred into motion, Sarnoff vowed consumers would purchase 75,000 color sets by the end of 1954. In this instance, his penchant for prediction failed. When only 5,000 sets sold, skeptics accused RCA of exaggerating their system's capabilities.[104] Other electronics firms scrapped plans to manufacture their own color sets, and RCA was left to champion color broadcasting on its own. Low sales over the next two years prompted *Time* to label color television "the most resounding industrial flop of 1956," but by then RCA had too much at stake to abandon the technology.[105] "We had our arms around color," Elmer Engstrom later remarked, "and we couldn't let go as the others could."[106]

More than their technological deficiencies, the convergence of the DSRC's shift to basic research and the divisions' focus on color television spelled doom for Nicoll, Kazan, and Rajchman's respective flat-panel projects. Both of the Magnalux prototypes required substantial development work, but the divisions saw no incentive to create a successor to the CRT given the time and money previously committed to color television. Rajchman's project came under particularly harsh scrutiny. "Everything we did, we had opposition from managers at Camden and various places," George Briggs explained. "They just wanted to have as much money as possible to be able to develop the tubes that were needed, the cathode ray tubes for color television."[107]

Nicoll and Kazan might have avoided such disputes over funding with their electroluminescent panels, which they now pitched primarily as medical instrumentation rather than CRT replacements. Unfortunately, tests of their radiology equipment at the University of Pennsylvania concluded that the "the quality of the picture was not good enough and the response of the panel was much too slow for clinical use."[108] With none of the pomp surrounding its presentation four years earlier, the DSRC's management canceled work on Nicoll and Kazan's light amplifier in 1960, consigning it to obscurity alongside Rajchman's mural television.[109]

THE LEGACY OF THE LIGHT AMPLIFIER(S)

Neither Sarnoff nor his lieutenants commented on the end of light amplification research at the DSRC. The General continued to harbor an interest in mural television, going so far as to install a conventional CRT in the walls of his Manhattan office, hidden behind a painting that slid away after dialing a specific combination of digits on his phone.[110] But after his golden anniversary, the color TV campaign dominated his private thoughts and public speeches. The competitive realities of the television market trumped the light amplifiers that Sarnoff once framed as television's future.

The use of the plural term *light amplifiers* in the previous sentence was deliberate. After all, between 1951 and 1956, neither Sarnoff nor RCA's technical staff treated Magnalux as a single, easily identifiable entity. The General had seized on the DSRC dedication ceremony to encourage investigations into areas his research managers deemed important to the company's long-term interests, but his later speeches suggest he had no clear sense of Magnalux's final form. For their part, the scientists and engineers who responded to his challenge acted to further their own research agendas rather than any preordained technological outcome. Epstein's experience working with theater television, for example, led him to construe Magnalux as a home projection system. Similarly, Nicoll and Kazan's panels emerged from fundamental investigations of electroluminescence and photoconductivity, while Rajchman viewed the call for a light amplifier as an opportunity to evaluate the relevance of computer addressing techniques to electronic displays.

Essentially the birthday gift speech served as a technological Rorschach test for RCA executives and engineers, an aspirational goal whose appearance shifted in response to each group's evolving understanding of light amplification.[111] The nascent state of electroluminescence research within the DSRC led Sarnoff to avoid overly specific descriptions of Magnalux's capabilities, but his call to action still encouraged personnel with a variety of disciplinary backgrounds to embrace the possibility of a television that could hang on the wall.

As Nicoll, Kazan, and later Rajchman blazed their respective trails toward that objective, their rising awareness of the underlying material logics of electroluminescent phosphors, photoconductive powders, and magnetic cores further altered how their superiors spoke about Magnalux. Discussions of projection systems gave way to light-amplifying panels before yielding to the ideal of a fully operational mural television. Sarnoff's public endorsement of the latter at his golden anniversary dinner, minutes after accepting Nicoll and Kazan's panels as the fulfillment of his vision, testifies to the power of both groups' ideas.

If the light-amplifier projects affirmed that RCA's technical staff could shape how management thought about technological innovation, it also showed that their influence had limits. Engstrom and Ewing's stewardship had granted scientists and engineers an unprecedented amount of autonomy within the walls of the Princeton labs, but that authority did not extend to other parts of the corporation. To the contrary, this freedom to engage in fundamental research, while critical to the incubation of electroluminescent display prototypes, only exacerbated tensions between the DSRC and RCA's other divisions.

As much as any technical factors, these institutional rivalries and the delayed payoff on color television spelled doom for both Magnalux initiatives. Compelling as Rajchman's prototype may have been, it was too cumbersome to mass produce and could not match the picture quality of existing TV sets. Kazan and Nicoll's panels survived slightly longer because of their creators' decision to pitch them as radiology equipment rather than CRT substitutes, but medical applications were a small piece of RCA's overall business. In the absence of compelling demand from the medical community, the company had few reasons to improve panel performance or fabrication processes.

While the quest for Magnalux never resulted in a commercial light amplifier, it should not be treated as an unmitigated disaster. Rather, it highlights the contingency of technological success and failure and the capacity of corporate researchers to blur the lines between those categories. RCA's executives and technical staff drew very different lessons from this first wave of flat-panel research. Viewed from the company's New York headquarters, the operating divisions' indifference to light amplification suggested that a reevaluation of the linear model might be in order. Perhaps it was time to rein in the fundamental explorations underway at the DSRC in favor of more direct participation in applications development.

Meanwhile, in Princeton, personnel built on the foundations that Nicoll, Kazan, and Rajchman had laid, confident that despite this shift in R & D strategy they could secure the necessary support to advance their research objectives. As new leaders implemented policies explicitly aimed at curtailing speculative projects, these researchers continued to pursue the General's dream of a flat-panel television long after the word *Magnalux* had vanished from RCA's collective memory.

2 A FUMBLING PRELUDE, 1956–1966

The average American in 1980 lived ten to fifteen years longer than his counterpart in the 1950s thanks to improved medical diagnostics and the elimination of polio, tuberculosis, and cancer. The atomic generators installed in his basement became commonplace well before the start of the new decade, allowing chemical engineers to devote their attention to finding new uses for fossil fuels. His mail was delivered by guided missiles, whose contrails stretched across an increasingly crowded sky filled with pilotless aircraft and fleets of personal helicopters. Admittedly, this rendered "the principal airways almost as busy as the highways on the ground," but electronic traffic control ensured safe transit in all directions.[1] Similar advances in automation increased workplace efficiency, granting employees in the nation's factories more leisure time for "appreciation of the better, and perhaps the best, in art, music, and letters."[2]

All of these technological and cultural advances, "unprecedented in kind and in volume," were possible thanks to an incipient revolution in electronics, at least according to David Sarnoff.[3] In a January 1955 article in *Fortune*, RCA's chairman provided a detailed vision of the fantastic world that lay in store as scientists unlocked the secrets of solid-state phenomena. To him no technology better represented "the fabulous dimensions of the teeming novelties now gestating in hundreds of laboratories, large and small" than the electroluminescence research underway at the DSRC.[4] While he supplied few details about RCA's preliminary attempts to create displays "freed from the prison of a vacuum bulb," Sarnoff swore that such work would make "obsolete the television tube of today, while bringing bigger and sharper pictures in color as well as in black and white."[5]

By the time of Sarnoff's golden anniversary dinner, it seemed his dream of a mural television might bear fruit before 1980. Yet the bright promise of his "fabulous future" dimmed somewhat during the latter half of the 1950s as RCA struggled to recoup its multimillion dollar investment in color television. Already facing pressure from the company's operating divisions to shift resources toward application development, DSRC researchers soon found themselves attacked on two additional fronts. First, they faced an external assault from the Eisenhower administration, whose vigorous pursuit of antitrust litigation threatened the corporation's traditional sources of research funding. In the midst of these legal battles, Sarnoff also appointed a new second-in-command, John Burns, who was committed to reining in the DSRC's fundamental research program.

From the perspective of personnel engaged in mural television projects, Burns's tenure was something of a mixed blessing. In response to a federal consent decree limiting RCA's ability to collect television patent royalties, Burns embraced a strategy of "profitable diversification," aimed at making RCA a leader in nonentertainment sectors such as military electronics, satellite systems, and, above all, digital computing.[6] These policies did not explicitly halt flat-panel display discussions in Princeton. Indeed, at least one DSRC researcher wrote at the time that "many people believe picture-on-the-wall displays represent a kind of ultimate toward which engineers should work."[7] All the same, Burns's regulations discouraged work on such an obviously long-term objective, and as the 1950s drew to a close, members of RCA's technical staff quietly shelved their flat-panel prototypes.

What Burns could not anticipate was that the institutional changes he implemented would supply RCA scientists and engineers with the resources to revive their mural television investigations following his 1961 departure from the company. His simultaneous cuts to the DSRC's budget and emphasis on defense-related projects inspired researchers to apply for government contracts to fund work that might otherwise have been deemed overly speculative. They found additional support from Burns's choice to serve as laboratory director, James Hillier, who saw mural television as a promising avenue to explore once color sets became profitable. Finally, Burns's interest in electronic data processing culminated in the formation of a new Computer Research Laboratory, which offered a site for display development separate from the company's established TV experts. When DSRC investigators observed the unique electro-optic properties of liquid crystals in the early 1960s, it would be from within this schizoid world, divided between television and computer researchers, that they would form a team to transform those new materials into displays.

CONSERVATIVE ATTACKS, RADICAL APPROACHES

Although Sarnoff chose to highlight electroluminescence in his *Fortune* article, his public enthusiasm for the subject and mural television research in general dwindled following the unveiling of Magnalux the following year. His silence on the matter did not translate into reduced interest in either topic among RCA's technical staff, however. On the contrary, by the end of 1957, they had outlined half a dozen possible methods to construct "a television reproducer thin enough to be hung on a wall like a picture," enough to merit an extended discussion in the DSRC's annual report.[8]

Managers in Princeton classified these efforts as "conservative" or "radical" depending on how each fulfilled the tasks of switching and illumination. Devices based on the "conservative attack" accomplished both functions using a guided electron beam, much like a traditional CRT-based television set. Switching occurred wherever electrons struck a screen, while modulating the beam's intensity to correspond with an incoming video signal ensured the associated phosphors emitted the proper amount of light. In contrast, the "radical approach" abandoned the use of cathode rays, treating a display like a "miniature telephone exchange in which 200,000 numbers are called in sequence to inform each one what the illumination should be for the next 1/30 second."[9]

From a corporate perspective, RCA possessed no strong attachment to either conservative or radical approaches toward display development so long as the resulting television could "equal or exceed cabinet-type [i.e., CRT] reproducers in economy, compactness, image size, brightness, and quality."[10] Consequently, research continued along both lines at the DSRC. Those with previous television systems expertise preferred to concentrate on conservative projects. Some focused on modifying CRTs—either by moving the source of the electron beam[11] or increasing its deflection angle[12]—to create more compact sets with larger screens. Such steps had to be taken carefully, because CRTs required an internal vacuum to ensure the free passage of electrons, and the greater surface area of large, flat screens raised the risk of implosion. Consequently, some, such as light amplifier coinventor Frederick Nicoll, preferred to consider how images on a CRT might instead be projected on a thin electroluminescent screen.[13]

Radical solutions to the mural television question attracted computer researchers, whose familiarity with matrix addressing offered a promising means of selectively switching the thousands of picture elements in a TV display. Jan Rajchman's transfluxor-driven prototype epitomized this approach. As Rajchman struggled to fabricate a model that matched the resolution of conventional televisions, other members of his group, including George Briggs, applied similar construction techniques to new projects, including an array of small "gaseous discharge elements," an early attempt to build a plasma display.[14]

Television prototypes constructed along both lines could generate moving images, but by early 1958 only modified CRTs left Princeton for evaluation at RCA's vacuum tube and television manufacturing divisions.[15] The DSRC's management acknowledged Rajchman and his team but noted that neither his mural television system nor any of its matrix-addressed descendants were "competitive with existing cabinet-type TV."[16] Further development was put on hold pending improved materials and fabrication techniques.

Rajchman tolerated this assessment without complaint. He had, after all, previously called attention to the same difficulties—inefficient phosphors, the high operating power of magnetic switches, and time-intensive construction processes—in his own reports.[17] Still, the actions he took after shelving his mural television prototype hint at the deeper economic and legal factors underlying the company's call to curtail development. Rajchman also had an additional incentive to set aside flat-panel projects after RCA's leadership turned to him to reinvigorate the DSRC's computer research program.

CHASING RAINBOWS

RCA's involvement in electronic computing dated to the late 1930s, but unlike IBM or Remington Rand, the firm initially preferred to position itself as a supplier of electrical components rather than complete computer systems.[18] The majority of this work was linked to military projects, though there were signs that the firm wished to expand into the civilian sphere. In March 1957, for example, RCA finished installing its first digital computer at the Army Ordnance Tank-Automotive Command in Detroit.[19] The $4.1 million vacuum-tube-based system was supposedly built to track replacement parts for combat vehicles, but its name—BIZMAC—revealed RCA's wider interest in "business machines."[20] Only six BIZMACs were ever sold, but as RCA's vice president for commercial electronics, A. L. Malcarney, told the *New York Times*, the company had "just scratched the surface" of the mainframe market.[21]

This renewed interest in computing coincided with the apparent collapse of David Sarnoff's color television campaign. The General had proclaimed that 1.5 million color sets would be in operation by the end of 1956, but the actual figure was closer to 75,000, leading critics such as Zenith president Eugene McDonald to accuse RCA of "premature tub thumping."[22] A combination of high prices, uneven performance, and the fact that color programming was limited to a handful of NBC shows left the public cold and RCA hard pressed to recover its $100 million investment in the technology.[23]

Sarnoff tried to assuage his shareholders, noting in a 1957 statement that the $7 million losses associated with color TV during the previous year were "inescapable for anyone who would pioneer and lead the way in a new field."[24] He urged the public to recall the start-up expenses required to bring black-and-white to the market, but these parallels failed to persuade them to embrace color. Other firms blamed the technical shortcomings of RCA television consoles, such as GE president Ralph Cordiner, who joked that "if you have a color set, you've almost got to have an engineer living in the house."[25] CBS vice president Richard Salant justified their network's resistance on financial grounds: "There's no public demand and no advertiser interest," Salant asserted in 1958. "Nobody gives a damn now. Suddenly, some day, color TV will blossom. We guessed wrong when we thought it would come much sooner."[26]

Though Sarnoff would be loath to admit it, the opposition to RCA during this period was partially self-induced, the result of intellectual property policies that alienated firms that might otherwise have entered the color television business. Perhaps as a result of its origins as a holding company overseeing the radio patents of GE, Westinghouse, AT&T, and the United Fruit Company, RCA was perennially worried about maintaining control over its own inventions.[27] After a brief attempt to restrict the distribution of RCA components to dealers who agreed to handle its entire product line, in 1927 the company adopted a new strategy known as package licensing. Under this arrangement, firms wishing to produce a piece of equipment covered by an RCA patent could buy a license granting them permission in exchange for royalties on all future sales. The term *package licensing* derived from the company's practice of requiring firms to purchase licenses for *all* of its patents even if only one was being used in a given product.[28]

As the RCA patent pool expanded, so too did resentment among its competitors, whose leaders found their business strategies increasingly subjugated to Sarnoff's. Two of these firms, Zenith and

Philco, took the matter to court, claiming that RCA's actions represented a deliberate attempt to monopolize the radio, television, and electronics industries, but the more damaging attack came from the federal government.[29] In February 1958, the Department of Justice filed what the *New York Times* referred to as "by far the most important criminal antitrust action in the five years of the Eisenhower Administration—economically and politically."[30] Eager to avoid bad publicity and the high costs of an extended lawsuit, RCA accepted a government offer to negotiate an out-of-court settlement. The resulting consent decree, announced in October 1958, marked the end of package licensing on "radio-purpose" (i.e., radio and television) electronics.[31] In addition to paying a $100,000 fine, the company agreed to "license on a royalty-free basis all its existing patents relating to the manufacture, use or sale of radio purpose apparatus" and "allow applicants to choose among these existing patents without taking the entire package."[32] RCA also placed one hundred of its color TV patents—along with related patents developed by half a dozen other companies—into a royalty-free pool to encourage industry-wide experimentation without placing RCA at a competitive disadvantage.[33]

Sarnoff considered the 1958 consent decree among the most humiliating incidents in his career. While he was spared the personal embarrassment of being named as defendant in a criminal suit, his company's exclusive control over color television was lost forever.[34] More troublingly, RCA found itself cut off from the patent royalties that had ensured the independence of its research facilities. The DSRC's 1954 shift toward "building-block" investigations had hinged on a revenue stream uncoupled from marketing deadlines or consumer applications. In the absence of those funds, the Princeton laboratories would face harsh budget cuts and difficult decisions about which projects to prioritize.

The dilemma confronting RCA would seem hauntingly familiar to its peers in the years to come. Whether they cared to admit it or not, several important American industrial laboratories relied on de facto monopolies to create a space where fundamental research could be conducted without agonizing over profit margins. Perhaps the most well known was the DSRC's New Jersey neighbor, Bell Labs, which was founded to help AT&T maintain the US telephone network. So long as the government sanctioned AT&T's near-total control over America's phone lines, the company could afford to sponsor the exploratory investigations that led to the transistor, solar panels, and many other technologies.[35] Once that protection was removed following the breakup of the Bell System in 1984, the competitive pressures of the market forced AT&T to narrow its research focus.[36]

RCA's reaction to this situation was to seek out a replacement for its lost licensing revenue, a search which led to a reexamination of electronic data processing. Computers, after all, were not "radio-purpose electronics" under the terms of the 1958 consent decree, which meant the company was free to collect royalties on its computer-related patents.[37] Before that could happen, the DSRC would have to undergo a transformation, scaling back its traditional emphasis on consumer electronics in order to kick-start its computing initiatives. Overseeing this strategic pivot was RCA's newly appointed president, John Burns.

"NO NEWCOMER TO RCA"

Since its incorporation in 1919, four men had served as RCA's president. Each possessed qualities that the company's board of directors deemed important to the firm's future. The first, Edward Nally, was formerly head of American Marconi and one of the few people with the technical and managerial skills to establish reliable wireless connections between the United States, Europe, and Asia.[38] Believing that RCA needed a more charismatic spokesman following the transition to broadcasting, in 1923 the board replaced Nally with James Harbord, a military hero who had served as chief of staff for General John Pershing—the leader of the American Expeditionary Force during the Great War.[39] Harbord's promotion to chairman in 1930 corresponded with RCA's successful break from its corporate shareholders, and he endorsed the architect of that split, David Sarnoff, as his successor.[40]

Together, Harbord and Sarnoff oversaw the expansion of RCA's manufacturing facilities and the drive to develop electronic television. When Sarnoff became chairman following Harbord's death in 1947, he selected Frank Folsom to fill his old job. Folsom's experience as a merchandiser for Montgomery Ward and RCA Victor was instrumental to the postwar boom in black-and-white TV sales, but the death of his wife and the intense legal wrangling over color television took a heavy personal toll.[41] Even before the Justice Department's antitrust intervention, it was clear that Sarnoff would need to replace Folsom. The question on everyone's lips at RCA was what sort of lieutenant the General would appoint in his stead.

The obvious candidate, particularly to those at the DSRC, was Elmer Engstrom, the director of RCA's laboratories, who had recently been promoted to executive vice president. His scientific qualifications were above reproach, and he had a strong working relationship with Sarnoff.[42] But as much as the General respected Engstrom, he decided that the soft-spoken scientist was not the dynamic leader RCA needed to maintain its place at the forefront of the electronics industry. Rather than draw from the company's existing pool of talent, he looked outside of RCA for Folsom's replacement.

In a January 1957 public statement, Sarnoff pointed out that John Burns was "no newcomer to RCA for he has been intimately associated with our activities for the past ten years."[43] A senior partner at the consulting firm of Booz, Allen & Hamilton, Burns was a nationally recognized management expert who had assisted Sarnoff and Folsom's postwar drive to reconstitute RCA as a large-scale manufacturing enterprise. He may have been an outsider, but he also possessed a doctorate in metallurgy from Harvard and had cut his teeth in the foundries of Republic Steel.[44]

The General expressed hope that this "unique combination of scientific and engineering knowledge, experience in industrial production and operation and an exceptionally broad understanding of business generally" would allow RCA "to keep pace with the changing demands and great opportunities of the rapidly expanding electronics industry."[45] Keeping pace with the industry meant first reinvigorating RCA's color television strategy so that its flagship product would no longer be a drain on its resources. Beyond that, Sarnoff and the board thought Burns's business connections would facilitate the company's expansion into sectors outside of home entertainment.[46]

For his part, Burns framed his arrival as an opportunity for RCA to reconsider its priorities. "RCA cannot be all things to all people," he cautioned, "but it must be in a leadership position in important things in our chosen fields. This means increased growth and effort, but it also points up the great need for selectivity and the timing of our efforts."[47] Color TV would remain the company's primary focus, and Justice Department officials hailed Burns's involvement in the 1958 consent decree negotiations as "a stroke of industrial statesmanship."[48] At the same time, the terms of that settlement rendered television research that did not directly increase RCA's market share an unprofitable investment.

Burns kept his eye on the bottom line and cited the consent decree to authorize a reevaluation of the firm's R & D policies. Soon after taking office, he convened a panel of RCA executives, high-ranking members of the technical staff, and representatives from the operating divisions for "a thorough review of all the projects under way in the research activity."[49] RCA scientists, who had previously acted with minimal oversight from the divisions, viewed this arrangement as an infringement upon their autonomy. In effect, these panels forced the Princeton laboratories to justify their previously unrestricted fundamental investigations on an annual basis.[50]

Burns's move met with applause from the divisions and outcry from the DSRC's theoretically inclined majority. To implement this new policy and ease tensions between the two groups, he appointed James Hillier as vice president in charge of RCA Laboratories. Hillier, a physicist and the mastermind behind America's first electron microscope, replaced "building block" advocate Douglas Ewing. Under his watch, fundamental investigations would not disappear from the DSRC, but such work would now bolster Burns's strategy of "broadening the base" to make RCA as formidable a force in defense and industry as it had been in consumer electronics.[51] Whenever possible, research in Princeton would be linked to products under development at the operating divisions. More significantly, Burns, acting "in accordance with the decision made by RCA management to participate vigorously in sales of data processing equipment," ordered Hillier to substantially increase the DSRC's computer research activities.[52]

During his time at RCA, no figure became as closely associated with the resurgence of commercial computing as Burns, whose consulting experience made him seem well suited to lead such a venture. Shortly before being tapped as president, he had overseen a major reorganization at IBM and recognized the difficulty of whittling down that company's 87 percent market share while continuing to invest in color television.[53] His solution to this dilemma was to develop sophisticated computer systems that were compatible with IBM software but could be sold at lower costs. In September 1957, six months after the delivery of BIZMAC, Burns negotiated a cross-licensing agreement with IBM president Thomas Watson Jr., and the following year, he announced the creation of the RCA 501: the company's first transistorized computer.[54]

RCA boasted that the 501 was "the world's most advanced electronic data processing system," but sales remained sporadic.[55] Burns admitted that the cost of entering the data-processing industry "was and is continuing enormous but so is the industry's potential."[56] He considered an investment in computing particularly sound given the shifting dynamics of Cold War science funding. "RCA not only staked out its share in one of the major electronics growth markets," Elmer Engstrom explained in an article a few years later, "but, equally important, it acquired the basis for continuing preeminence in other electronics fields where computers and computer systems were indispensable to progress."[57]

Specifically, the growth of RCA's data-processing activities left it poised to benefit from the outpouring of government communications and computing contracts that followed the launch of Sputnik in October 1957. Eager to make up for their lost patent revenues, DSRC research managers enthusiastically welcomed federal funding, justifying such work as "sufficiently fundamental in character that it cannot fail to have important significance in our long-range commercial programs."[58] As a result of Burns's policies, by 1960, more than half of the research conducted in RCA's laboratories could trace its funding to government agencies.[59]

MURAL TELEVISION UNDER BURNS

The press hailed Burns as "management's Renaissance man" and a champion of "corporate versatility," but the DSRC's growing focus on data processing did not bode well for the company's flat-panel display projects.[60] Given the technological similarity between the magnetic addressing circuitry used in "radical" television prototypes and the core memories found in mainframes, personnel associated with the former were among the first reassigned to the DSRC's computer group. Leading the way was none other than Jan Rajchman, who dispensed with his electroluminescent displays to manage RCA's involvement in a US Navy-sponsored computing initiative. The $25 million effort, code-named Project Lightning, sought to create a 1 gigahertz processor, several orders of magnitude faster than commercially available computers.[61]

Although RCA's electronic data-processing division in Camden would deliver the completed system, the DSRC assumed primary responsibility during Project Lightning's early phases. This division of labor reflected the fact that the navy contract stipulated that the new computer could not rely on existing transistor technology.[62] Instead, between 1957 and 1962, researchers in Princeton would have to create entirely new logic and memory circuits for "ultrafast" computing using relatively unexplored devices such as tunnel diodes, microwave oscillators, and superconductors.[63] Project Lightning, along with other defense-related computing projects, soon dominated the research agendas of Rajchman and his colleagues, leaving radical approaches to mural television to fall by the wayside.[64]

Conservative flat-panel investigations had a slightly longer life span in the wake of Burns's strategic realignment, but one by one, each of the CRT-based projects under consideration before 1957 folded. No matter how one positioned the electron beam's source or altered its deflection angle, modified CRTs could not consistently match the performance of existing picture tubes. The growth of computing activities and RCA's color crusade placed additional constraints on the DSRC's research budget, which made it harder to sustain the search for a wall-mounted television. In the DSRC's annual reports, previously distinct sections on "mural television," "thin cathode-ray tubes," and "electroluminescent displays" disappeared after 1960, subsumed within broader discussions of picture tubes or phosphor chemistry.[65]

At no point did Burns, or anyone outside of Princeton, declare that RCA had abandoned flat-panel television research. Rather, as RCA's engineers were conceding that "such a device still stands as a challenge today—a very difficult problem," its marketing staff was encouraging consumers to keep dreaming of a future filled with wall-mounted screens.[66] "How'd you like to hang your TV picture on the wall?" a 1959 supplement in *Popular Science* asked. "Although it may be several years before you can literally do that, with RCA's new line of 'Mural-TV' sets you can easily get the same effect."[67] Perhaps taking a cue from David Sarnoff's New York office, RCA offered to install custom cabinets for standard black-and-white or color sets, creating the illusion of a flat-panel display (fig. 2.1). "The ideal display of television today is as a picture-on-the-wall," one sales executive explained in RCA's *Electronic Age* (formerly *Radio Age*) magazine. "No matter how it is installed, a 'Mural TV' installation adds to the room's warmth and informality."[68]

Such assurances were not only restricted to the printed page. At a May 1961 press conference, RCA executives showed reporters life-size mock-ups of televisions that might be available for purchase in the 1970s, including a "large-screen color television console, less than five-inches in depth."[69] Among the speakers present at this event was

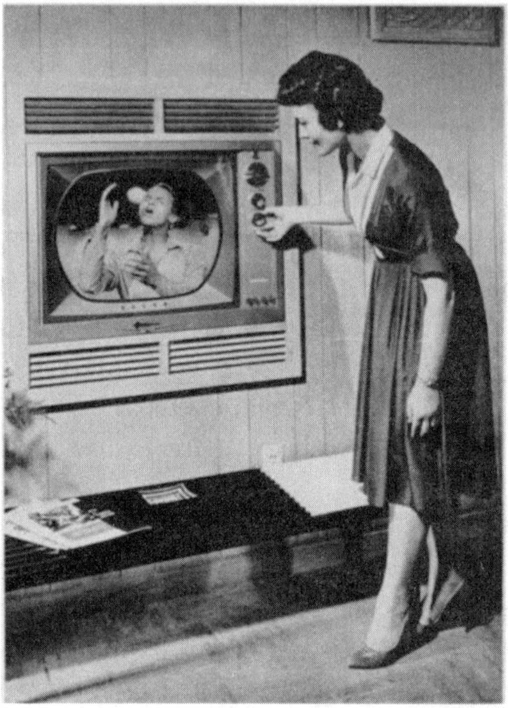

Figure 2.1. The inability of DSRC engineers to develop a flat-panel replacement for the CRT did not prevent RCA from capitalizing on the allure of "picture-on-the-wall" television. ("Built-In Entertainment," *Electronic Age* 19, no. 2 [Spring 1960], 10. David Sarnoff Library Collection, courtesy of Hagley Museum and Library. Scan courtesy of Linda Hall Library of Science, Engineering & Technology.)

John Burns, who gamely posed next to a model mural television despite having essentially terminated all flat-panel R & D projects at the DSRC (fig. 2.2).

"NOT A REVOLUTIONIST"

As Burns chatted with the press about what changes were in store for television during the 1970s, his future at RCA appeared secure. In a *Wall Street Journal* interview the following month, he expressed confidence that the company's earnings, which had dropped sharply in 1960, would recover by the end of the year. With one exception, all of

Figure 2.2. At a 1961 press conference, RCA executive vice president W. Walter Watts *(left)* and president John Burns *(right)* presented several mock-up television sets that RCA predicted would be available to consumers in the 1970s, including these flat-screen models. (Courtesy of RCA.)

RCA's divisions were operating in the black.[70] Even color television, which had hung like a $130 million millstone around RCA's neck, had started generating seven-figure profits, large enough to convince longtime holdouts such as Zenith to set aside their resistance.[71]

Electronic computing continued to sap the company's finances, but Burns asserted this was a natural result of relying on long-term rental income as opposed to the outright sale of data-processing equipment. “We are a major factor now in the computer business,” he declared. “Our program to cross over from investment to firm and mounting profits is on target.”[72] Concerns over computing forced Sarnoff to intercede on Burns’s behalf at a 1961 shareholders meeting. “Burns is not a revolutionist. His contribution to this company is not measured by figures,” the General explained. “He’s an efficiency expert of the first order. . . . I’m satisfied with him.”[73]

Yet such expressions of confidence masked Sarnoff’s unease about his second-in-command’s performance. Burns could claim several successes during his presidency, including the relocation of RCA’s consumer products operation to Indianapolis and the firm’s involvement in prominent defense projects such as the Ballistic Missile Early Warning System (BMEWS) and TIROS (Television Infrared Observation Satellite)—the world’s first weather satellite.[74] Nevertheless, his management style had provoked resistance throughout the corporation. Some opposition might have been expected at the Princeton labs, where Hillier warned against overemphasizing applications at the cost of long-term research.[75] But frustration was also evident within the operating divisions that were the supposed beneficiaries of Burns’s strategy. Personnel at the Industrial Electronic Products (IEP) division in Camden, where the RCA 501 was developed, complained of sudden, unproductive policy changes and long-distance micromanagement. As George Brown, the head of IEP at the time, reflected in his autobiography, “Burns was a very affable person and would perhaps have been a success if he had listened more and talked less.”[76]

Sarnoff might have overlooked these complaints if Burns’s ideas had generated higher earnings, but RCA’s data-processing projects increasingly seemed an expensive diversion from its core consumer electronics business. In an attempt to move past the RCA 501’s weak sales, Burns had authorized the production of two more transistorized computer systems. The smaller RCA 301, intended for businesses with basic data-processing needs, enjoyed modest success, but as one IEP executive noted, the larger 601 model suffered from “severe technical problems, both in a functional and in a manufacturing sense, and there were also severe financial problems, so much so that the company began to look for a way out of the program."[77]

The 601's problems left RCA with a serious blemish on its reputation and without a high-powered model to anchor its line of mainframes. Burns again dismissed such setbacks, arguing, as he had for years, that the corporation's future rested with three C's: computers, controls, and communications.[78] Unfortunately, this credo neglected what David Sarnoff considered the most important "C" of all: color television. After the chairman took Burns to task for this omission at a board meeting, George Brown informed his wife that the company's president "would not be with us much longer."[79] The General could forgive many things, but forgetting color television was an unpardonable sin.

Whether Sarnoff was genuinely offended by Burns's three C's remark, was disappointed in RCA's lackluster entry into commercial computing, or was responding to complaints from the company's divisions, Brown's prediction soon came to pass. In November 1961, the board of directors stripped Burns of most operational responsibilities. While he continued to manage NBC, supervision of the electronic data-processing, home instrument, and semiconductor divisions passed to Elmer Engstrom. A month later, Burns announced his resignation "for personal reasons," and Engstrom, who had been passed over in RCA's presidential search five years earlier, was tapped to take his place.[80]

As they had with Burns, the press cited Engstrom's combination of technical expertise and administrative ability as qualifications for the position, but many RCA personnel saw a crucial difference between the two men.[81] To them Burns was an outsider whose top-down call to transform the corporation's core business reflected a deep misunderstanding of its R & D procedures. Engstrom, by contrast, started his career as an electrical engineer working on broadcasting equipment and remained on good terms with personnel in Princeton and Camden as he climbed the corporate ladder.[82]

The replacement of the assertive, aggressive Burns with the

precise, professorial Engstrom signaled a partial reversion toward the company's earlier adherence to the linear model. Electronic data processing would remain a priority, but Engstrom phased out the 501 and 601 product lines to focus on the smaller, more popular RCA 301.[83] DSRC researchers meanwhile continued to consult with the operating divisions, but Hiller made it clear that such activities should not eclipse the laboratories' fundamental research programs. As the research center's 1962 annual report observed, "the increased diversification of product lines"—the hallmark of Burns's administration—"required that the search for innovation be directed increasingly to the building blocks of our products, the materials and active components, rather than to the products themselves."[84]

Even as RCA sought to distance itself from Burns, his tenure continued to shape research activities in Princeton. The weight he placed on government contracts as an alternative source of revenue, for example, persisted at the DSRC well after the influx of color television sales supplanted reliance on domestic patent royalties as a source of research funding. The growth of the research center's computer group also continued unabated once Burns left. And although he continued to encourage partnerships with other divisions, James Hillier was more willing than Burns to sponsor speculative projects at the DSRC, particularly if that work might prove relevant to RCA's consumer electronics business.

That description certainly applied to mural TV research, which resumed at the DSRC shortly after Burns's departure. Thanks to him, RCA scientists and engineers now had access to an assortment of new institutional and financial structures to support either conservative or radical approaches to flat-panel development. The success of these efforts would hinge on how well individual researchers could harness those resources in pursuit of a television that could hang on the wall.

PRACTICAL GUYS AND PHDs

The obvious venue for a mural TV revival was the DSRC's Systems Research Laboratory (SRL), which had previously overseen the development of RCA's compatible color system. By the time Burns left in 1961, the SRL had grown bloated and unwieldy, with a staff of over one hundred researchers assigned not only to radio and television projects but also "the development of new systems involving electronic techniques" ranging from radar antennae to computer memories.[85] This diverse portfolio might have appealed to a polymath like Rajchman, who transferred to the SRL in 1955, but Hillier viewed the lab's large size and diffuse focus as problematic given TV's centrality to the RCA brand.[86] He therefore took it upon himself to streamline the SRL and restore its traditional emphasis on television.

In September 1961, Hillier—acting with Burns's blessing—siphoned off a sizable portion of the SRL's staff to establish a new Computer Research Laboratory (CRL), leaving the remaining scientists and engineers to focus on "improving the quality and decreasing the cost and size of RCA color television receivers."[87] Early on, the newly reduced SRL dedicated itself to the synthesis of new phosphors and electron beam deflection studies reminiscent of the conservative CRT work from a few years earlier, but that changed at the end of 1962. At that time, growing competition in the color TV market spurred SRL director Allen Barco to once again call on RCA researchers to "investigate and evolve successors to the present type of cathode ray displays."[88] The goals of this SRL project were similar to RCA's previous flat-panel programs, namely, the creation of "low cost, thin, flat, lightweight displays with high brightness, contrast, resolution, and speed."[89]

Barco delegated responsibility for flat-panel display development to two SRL group leaders: Jay Brandinger and David Kleitman. Both had arrived recently in Princeton, and their technical backgrounds and managerial styles were quite different. Brandinger had spent eight years solving radio reception problems at an RCA research facility on Long Island before transferring to the DSRC in 1959.[90] His display group reflected his strong practical bent and included several of the company's most experienced television systems engineers. Ray Kell, for example, had spearheaded some of RCA's earliest color TV experiments in the 1940s, while Dalton Pritchard was part of the team that determined how to transmit a color signal within the bandwidth that the FCC allotted for black-and-white television.[91]

Where Brandinger, an engineer, treated the development of a flat display as a task best solved by established CRT experts, Kleitman, a solid-state physicist who joined RCA's staff in 1957, latched on to Barco's directive to investigate "basic phenomena and techniques that can be developed and used in the new displays."[92] He received permission to hire a cohort of newly minted PhDs to consider the problem from a wide range of disciplinary perspectives. One of those recruits, John van Raalte, arrived in Princeton in 1964 after earning degrees in electrical engineering and solid-state physics from MIT. He immediately noticed a difference between Kleitman and Brandinger's teams:

> I think the people that I call the "practical guys," you know the Dalton Pritchards and Ray Kells, knew intuitively what to work on and had good ideas and drove toward success or whatever, and nobody really knew how to manage a bunch of young Ph.D.'s. . . . [Istvan] Gorog had done something on plasma physics. Mike Kaplan was an organic chemist. You know, at the time,[it was] not at all clear how he would fit into a display group.[93]

Brandinger and Kleitman's research groups provided an apparent test case to determine which of the DSRC's constituencies could better tackle the consumer electronics challenges of the 1960s. On one side were the "practical guys," veteran researchers who often possessed limited formal education but a great deal of workbench experience; on the other were scientific specialists with advanced degrees but no familiarity incorporating their findings into commercial products. In the end, neither group succeeded in developing a market-ready flat-panel display, but watching how each approached the task revealed two things. First, both groups treated the project as exploratory research whose near-total reliance on internal funding was reminiscent of previous building-block projects. Second, regardless of disciplinary background, SRL researchers demonstrated a common affinity toward conservative displays and clung to the same fundamental addressing mechanism: a beam of particles sweeping across a screen.

Brandinger's group believed that the key to RCA's future flat-panel ventures could be found by looking at recent advances in projection technology, specifically a Swiss invention known as the Eidophor. At the heart of an Eidophor projector was a mirror covered with a thin layer of oil. Firing an electron beam at the mirror deposited charge on the oil, causing surface deformations. If a bright light shone on the mirror, the resulting reflection could be thrown on to a screen, with special optics transforming the distortion patterns into a moving television image. Early Eidophor systems could only generate monochrome images, but by 1958 GE had constructed a full-color projector.[94]

When combined with a powerful enough light source, the Eidophor could produce very large, very bright images, but the oil used in such systems tended to degrade over time, limiting its operational life span.[95] Kell and Pritchard each sought to overcome this deficiency. Kell replaced the Eidophor's oil-coated mirror with a deformable aluminum film, which he mounted on the front of a conventional CRT.[96] Pritchard's setup eliminated the mirror altogether. Instead, he fired the electron beam at a crystal whose ability to rotate polarized light depended on the charge applied to its surface (fig. 2.3).[97] Both of these systems proved capable of producing television images but only in monochrome. In addition, each relied on the image-forming targets, which were not amenable to mass fabrication.[98]

As Brandinger's team concentrated on CRT-based projection systems, Kleitman's group was, in Van Raalte's words, "more focused on brand new technologies and finding out what they might have to offer for displays."[99] The most compelling of these new technologies was the laser. A DSRC quantum electronics group had examined the use of lasers in communication systems since 1960, but Kleitman encouraged further consideration of their display potential.[100] The combination of solid-state physics, electrical engineering, and optics involved in laser construction also made them attractive to a team

Figure 2.3. Researchers in the DSRC's System Research Laboratory believed CRT-based projection systems, such as this one developed by Dalton Pritchard, offered the most effective means of creating a flat-panel television display. (RCA Laboratories, *Research Report 1965*, 31. David Sarnoff Library Collection, courtesy of Hagley Museum and Library.)

of newly minted PhDs looking for a cutting-edge project. By 1965, the group had found several ways to modulate the wavelength, and consequently the color, of laser light. Such techniques hinted at the prospect of high-resolution "laser beam projection displays," but before the team assembled a prototype, several members set aside TV research to explore holography.[101]

The decision to abandon laser projection represented an admission of its technical difficulty and the fundamental limits that particle beam scanning, whether it involved electrons or photons, imposed on display design. Given the risk of implosion as a CRT's screen size increased, the closest either SRL group could come to mimicking its performance while adhering to a conservative addressing scheme was a projection system. Kleitman authorized a few tentative forays beyond the beam in search of a self-contained panel television that would "bridge the gap between vacuum and solid-state displays."[102]

In spite of his claims that such "hybrid" technologies were an ongoing priority, the SRL's flat-panel research ended in 1966, as did the laboratory itself.

Before the end of that year, Hillier consolidated the SRL and parts of the DSRC's Acoustical and Electromechanical Laboratory into a new Consumer Electronics Research Laboratory (CERL) in order to prevent a complete relapse into the building-block mind-set of the 1950s. The CERL's "responsibility for research that will generate new and/or improved electronic services for the consumer" emphasized improving existing TV systems over the creation of new displays.[103] Work continued on the CRT devices Brandinger nurtured to the prototype stage, but solid-state displays receded from the CERL's research docket, leaving such work to be pursued not by television experts but the engineers associated with RCA's Computer Research Laboratory.[104]

WHERE LIGHTNING STRUCK

The establishment of the Computer Research Lab coincided with the 1961 conclusion of DSRC involvement in Project Lightning. After developing high-speed computing circuitry for five years, Rajchman and his CRL colleagues now had to formulate new projects that would both legitimize and benefit from their newly acquired institutional independence. One area ripe for potential involvement was the improvement of RCA computer peripherals—input and output devices such as printers and card readers, which had previously been obtained from outside vendors as a cost-saving measure. The unreliability of this equipment had contributed to customer dissatisfaction with the RCA 301 and 601 mainframes, so in 1962 the firm decided to finance an internal peripherals R & D program.[105] The post-Lightning CRL would therefore expand its previous focus on memory systems and processors to include such diverse topics as optical character recognition, printer mechanisms, and display devices.[106]

Presumably, some of the CRL's interest in displays could be traced back to Rajchman, who viewed the manpower and money that Burns funneled into computer research as a means to resurrect his old mural television prototypes. This time around he was determined to avoid the mistakes that had doomed his earlier efforts, most notably the high power needed to activate the magnetic transfluxors behind each electroluminescent picture element. He therefore assigned the task of devising a more efficient form of addressing circuitry to an electrical engineer named Bernard Lechner. Lechner had joined RCA in June 1957 expecting to work on television systems. As he recalled, instead he was assigned to Project Lightning despite the fact that he "didn't really know anything about digital circuitry or digital applications of magnetic devices."[107] Still, he retained an interest in displays, and after several years tinkering with logic circuits, he was excited to inherit Rajchman's old project.[108]

Lechner began by conducting a literature review in RCA's technical library. He soon discovered that during the DSRC's flat-panel hiatus, several people had published articles suggesting that the power requirements of a matrix-addressed display could be sharply reduced by replacing magnetic cores with ferroelectric switches.[109] Ferroelectric materials' behavior under an electric field was similar to ferromagnetic materials—like iron and nickel—under a magnetic field. In both cases, the substance formed field-aligned regions called "domains," whose polarity could be switched to correspond to a 1 or a 0 in binary code. The main advantage of ferroelectric materials in this case was that their impedance (a measure of electrical resistance in AC circuits) was much closer to that of Rajchman's electroluminescent phosphors, theoretically offering a significant reduction in the energy needed to produce a bright picture.

On Rajchman's recommendation, Lechner familiarized himself with the work of Ennio Fatuzzo, a researcher at RCA's recently established laboratory in Zurich.[110] Fatuzzo had immersed himself in studies of ferroelectric switching with an eye toward creating what he termed a "FE-EL [ferroelectric-electroluminescent] Mural TV."[111] Rajchman suspected Fatuzzo's research might enable the construction of a display based on the "transcharger," a ferroelectric analog to the ferromagnetic transfluxor, capable of performing the same functions at a much lower power.

The transcharger concept was not new. Rajchman had used the term in his mural television reports, and he and Briggs referred to it in a 1955 patent.[112] Later, Catholic University professor Charles Pulvari independently confirmed the utility of such a device in "large screen display devices with storage properties."[113] Lechner knew that these largely theoretical musings were hardly sufficient to enable construction of a functional display. "This all makes sense—a large flat panel—only if you could integrate the structure so that you could mass produce it at a reasonable cost," Lechner clarified in a 2009 interview. "So the question was how do you make a flat panel that would do that?"[114] Transchargers offered one solution, but they came in a variety of configurations, and it was unclear which would work best in displays, much less how to produce the 1,200 needed to match the performance of Rajchman's prototype. Nor was it possible, solely on the basis of Fatuzzo's research, to know which ferroelectric compound possessed the right combination of electrical properties for matrix addressing.

Unlike the SRL scientists, who chose to rely on internal funding, Lechner decided the best way to answer these questions was to obtain external support. With Rajchman's encouragement, he obtained a contract from Wright-Patterson Air Force Base to "develop simple and inexpensive solid-state control circuitry capable of meeting future Air Force solid-state display needs."[115] Once the money was in place, Lechner assembled "the beginnings of the nucleus of what became a group, although I didn't have the title of group head."[116] There was no shortage of skilled engineers in the CRL, but most of RCA's research on ferroelectric materials was being carried out in Zurich. Fortunately, Rajchman recognized this gap and recruited George Taylor, an Australian who had just completed a dissertation on ferroelectric computer memories at the University of London.

Taylor traveled to New York in the summer of 1962 and swiftly took on a central role in the CRL display project. His responsibilities included not only determining the chemical composition of the transchargers but also designing procedures to produce them in large numbers. By December 1962, he had decided to utilize ferroelectric ceramics rather than the crystalline materials from his thesis research, writing in his notebook that the former "have the advantage that they can be made in relatively large sheets which are suited for the mass fabrication of transchargers required for the display system."[117] Over the next year, Taylor studied "zillions of different variations in composition," before settling on a niobium-doped, lead-based ceramic, which could be easily molded into uniform strips.[118]

Meanwhile, Lechner partnered with fellow CRL engineers Anatole Samusenko and Juri Tults to assemble a series of increasingly sophisticated transcharger-driven displays. By the spring of 1965, they had built a seven-segment numeric indicator, two thirty-six-element "exercisers" (regular and high resolution), and a 120-element model capable of displaying moving images.[119] These proofs of concept persuaded the air force to sponsor the completion of a 1,200-element display. Much like Rajchman's prototype, it would consist of thirty rows, each containing forty picture elements. Rather than threading wires through hundreds of magnetic cores as Rajchman had in 1955, Lechner's team took advantage of evaporation techniques used in semiconductor manufacturing to deposit the metal electrodes for each transcharger directly on to Taylor's ceramic strips. To further simplify assembly, their display disposed of individual electroluminescent cells in favor of a single pane of glass sprayed with a green phosphor coating. "So we had two clear steps towards integration that he [i.e., Rajchman] didn't have," Lechner explained. "First of all, the EL [electroluminescent] panel was fully integrated—one piece. And secondly, the transchargers were integrated to the extent of having a half row at least on one piece of ceramic material rather than the individual cores with wire wound through them by hand."[120]

"The complete system was first operated on June 28, 1966," Juri Tults recorded in his lab notebook. "Static as well as moving images picked up by the TV camera were displayed on the display panel."[121] The ten-inch diagonal FE-EL prototype was smaller than Rajchman's

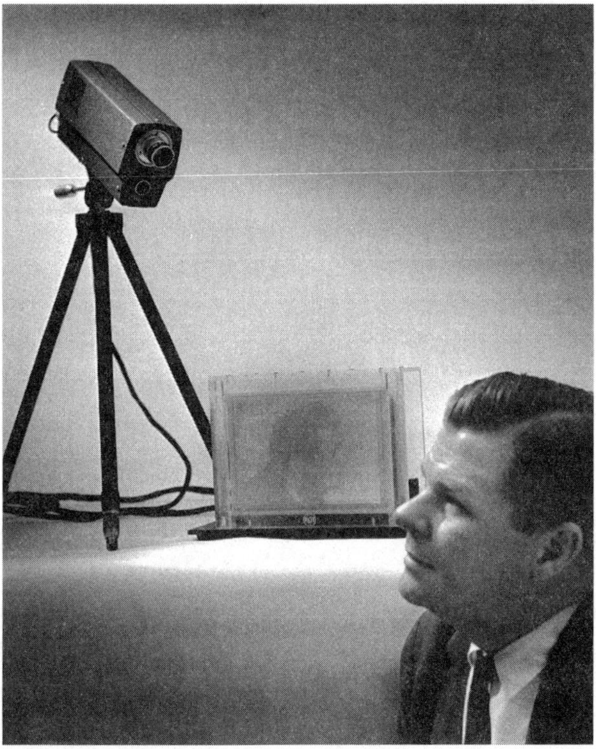

Figure 2.4. Although smaller than Jan Rajchman's mural television prototype, the Computer Research Laboratory's ferroelectric-electroluminescent display had a noticeably higher resolution, as seen in this photograph featuring project leader Bernard Lechner. (David Sarnoff Library Collection, courtesy of Hagley Museum and Library.)

twenty-three-inch model, but the larger number of picture elements per inch resulted in higher-resolution images.[122] The group reported that "the results have been surprisingly good considering the resolving power of the display is severely limited by only having 1200 elements. It was found, for example, that as many as 15 words could be read on the screen. Also features were sufficiently detailed to enable human faces to be identified" (fig. 2.4).[123]

Following demonstrations for RCA's top managers, including David Sarnoff, Lechner and Tults took an overnight train to Dayton to deliver their prototype to the US Air Force.[124] They boasted in their final contract report that theirs was "the most advanced display of its kind that has been built to date."[125] Of course there was still room for improvement—the transchargers were at best an intermediate step toward fully integrated circuitry, and there was some variation in the brightness of picture elements—but the CRL team promised their patrons that they could handle these difficulties. In the end, despite these assurances, the air force chose not to renew its contract, and without military funding the CRL display project drew to a close at the end of 1966. "Board displays several feet in size are practical for special applications through the FE-EL approach," the DSRC's annual report concluded, "but for smaller screens the system does not appear economical."[126]

MOBILIZING RESOURCES IN CORPORATE LABORATORIES

"There is no element of material progress we know today—in the biological and chemical fields, in atomics and electronics, in engineering and physics—that will not seem, from the vantage point of 1980, a fumbling prelude."[127] When Sarnoff wrote those words in his 1955 *Fortune* article, he had singled out electroluminescent displays as the embodiment of the "fabulous future" awaiting Americans in the next quarter century. Yet a decade later, the CRT still reigned supreme, and RCA was no closer to commercializing a flat-panel display based on electroluminescence or any other technology. Viewed in this light, Sarnoff's phrase—"a fumbling prelude" seems an apt description for the period between 1955 and 1966, at least so far as mural television is concerned.

Such a characterization, however, does a disservice to the DSRC scientists and engineers, who demonstrated a remarkable capacity to adapt to a rapidly changing research environment. Until the late 1950s, RCA's technical staff had internalized the assumptions of the linear model. They presumed that investigations into fundamental physical phenomena, conducted with no clear commercial objective in mind, would generate new patentable technologies. The licensing revenue derived from those technologies would then be siphoned back into the laboratories to support further research, which would result in more patents, and so on. The slow commercialization of color television and, more importantly, the Justice Department's antitrust suit interrupted this virtuous cycle and jeopardized the corporation's capacity for future innovation.

John Burns had participated in negotiations surrounding the 1958 consent decree well aware of its implications for RCA's research activities. Controversial as the decision to enter the computing market may have been, it was rooted in the belief that the company should prioritize technologies that might replace its lost radio and television royalties. While he was overzealous in his insistence that the DSRC focus on short-term application development, there was perhaps a modicum of truth in Sarnoff's assertion that Burns was "not a revolutionist."

At the same time, the R & D infrastructure that Burns left behind bore striking differences from the building-block days of the 1950s. Although Hillier was more open to exploratory research than Burns, he refused to allow a complete reversion to the DSRC's old country club mind-set. The laboratories would continue to collaborate with the operating divisions, particularly in the field of electronic data processing. Members of the technical staff could still carry out fundamental investigations, but they now knew it was in their interest to secure external funding to support that work.

Burns's policies terminated the first wave of mural television research at the DSRC, but they also, quite inadvertently, set the stage for a renewed attack on the problem after his departure. The relative success of the CRL team—whose prototype was the only flat-panel display developed during this period to be delivered to an external client—can be attributed to its ability to mobilize those resources more effectively than their SRL counterparts. Capitalizing on the CRL's newfound freedom, Lechner and his fellow engineers focused on a single project and obtained military funding to improve fabrication techniques and sustain fundamental investigations into ferroelectric ceramics and electroluminescent phosphors. The SRL, meanwhile, relied purely on internal funds and divided its energies among a variety of projection systems with only limited regard for the manufacturing difficulties involved.

Neither the SRL nor CRL succeeded in creating a commercially viable replacement for the cathode-ray tube, but the persistence of the mural television ideal testifies to the capacity of DSRC staff members to advance their own research agendas. Subverting the very structures Burns imposed to push RCA away from such speculative projects, they laid a foundation for flat-panel display research in Princeton that would last until the early 1970s. It was, after all, because of his experiences in the CRL that Lechner first learned about a new push to develop displays using liquid crystals, a project that, as he put it, "turned out to be much more promising."[128]

3 SCATTERED ORIGINS, 1961–1968

In December 1964, David Kleitman invited two staff members from RCA's Systems Research Laboratory (SRL) to witness an experiment. Over the past year, chemist Lucian Barton and engineer Warren Moles had become affiliated with Kleitman's flat-panel display group, revisiting Kazan and Nicoll's work on photoconductivity and electroluminescence.[1] Now they joined Kleitman and SRL leader Allen Barco to watch George Heilmeier, a relative newcomer to the David Sarnoff Research Center (DSRC), demonstrate the results of his investigations into an obscure class of materials known as liquid crystals.

The group watched as Heilmeier placed an orange mixture between two small glass slides, positioned the sandwich on a heated microscope stage, and shone polarized light through it. After connecting the sample to a power supply, he steadily ramped up the voltage across the mixture, and, as Barton recorded in his laboratory notebook, the material "changed color from orange to red in a quite uniform manner what really surprised me."[2] Kleitman and Barco, always on the lookout for new display technologies, were equally impressed, and, Barton wrote, "they want to consider liquid crystals as first priority problem to work out."[3]

Shortly after that meeting, Heilmeier found himself on the receiving end of a summons. News of his color-changing crystals had spread throughout the DSRC, eventually reaching the desk of Vladimir Zworykin, the inventor whose iconoscope camera and kinescope picture tube had allowed RCA to enter the field of electronic television. Zworykin had retired from active duty a decade earlier but stayed on site in Princeton as a technical consultant and honorary vice president. Though he had left television research behind to focus on medical electronics, Heilmeier's liquid crystals piqued his interest.[4]

Once again, the younger engineer presented his findings, noting that the mixture consisted of a "guest" dye dissolved in a liquid crystal "host" and that the solution's color changed under an electric field. Zworykin then asked Heilmeier about the research leading to this discovery.

> I explained to him how I had *stumbled* on the guest-host color switching effect. I'll never forget his reflective reply: "Stumbled perhaps, but to stumble, one must be moving." We were moving. It was only the beginning.[5]

Indeed, Heilmeier's successful demonstration of guest-host color switching persuaded Barco and Kleitman that liquid crystals merited further examination and led in short order to the formation of a research team dedicated to evaluating their technological potential.

Some members of this group would later frame the guest-host effect as the starting point for LCD development at RCA, but accepting this claim is problematic for two reasons.[6] The first relates to Zworykin's inquiry into the origins of Heilmeier's interest in liquid crystals. Contrary to his modest reply, the observation of electronic color switching did not result from haphazard stumbling but rather a series of methodical decisions that Heilmeier made in consultation with chemists and technicians. One of those researchers, a physical chemist named Richard Williams, had even filed a patent outlining how liquid crystals' unique electro-optic properties might be utilized in displays.[7]

More significantly, despite serving as the catalyst for RCA's liquid crystal program, within six months Heilmeier's team had turned their attention from guest-host displays to another electro-optic phenomenon. This new effect enabled certain transparent liquid crystals to take on a milky-white appearance when subjected to an electric field without the addition of dyes or reliance on polarized light. It was this "dynamic scattering" mode, not electronic color switching, that enabled the construction of the first practical LCDs.

Understanding how liquid crystals were transformed from a scientific curiosity into the basis for a new form of display requires us to reconstruct the investigative pathways that led Heilmeier first to the guest-host effect and later to dynamic scattering. It also necessitates consideration of the process through which he organized the team that incorporated dynamic scattering into functional prototypes in contrast to Williams, who had proposed a similar idea before setting it aside because of a perceived lack of support. Ultimately, the divergent outcomes of Williams's and Heilmeier's LCD projects depended on their conceptions of the industrial research enterprise and each man's ability to recruit allies from within RCA's management and among different laboratories at the DSRC.

LOOKING FOR SOME OF THE ACTION

Three years before his conversation with Zworykin, George Heilmeier had been a PhD candidate in search of a dissertation, completely unaware of the liquid crystals that would one day cement his reputation. The only child of a working-class family living in the Mayfair section of Philadelphia, Heilmeier had received a scholarship from the University of Pennsylvania and earned a bachelor's degree in electrical engineering in 1958.[8] Following graduation, he joined the technical staff at the DSRC and took advantage of RCA's advanced study program to pursue a doctorate from Princeton. By the spring of 1961, he had completed a master's thesis on solid-state microwave amplifiers, but he found the thought of expanding this topic into a dissertation unappealing.[9] "The competition for new ideas was getting tougher," he explained, "and, having had a taste of exciting research, I wasn't particularly interested in slugging it out on a mundane problem simply to get my degree."[10]

As a result, Heilmeier omitted microwave devices from a list of possible dissertation topics that he stapled in the front of his laboratory notebook.[11] The majority of these ideas revolved around semiconductors such as germanium and silicon, but after consulting with his faculty advisor and RCA mentors, he "left the relatively safe world of inorganic materials behind" to study phthalocyanine—an organic substance best known for its use as a blue pigment.[12] Half a century later, carbon-based compounds like phthalocyanine would become essential to a variety of high-tech applications, ranging from low-cost solar panels to energy-efficient displays based on OLEDs (organic light-emitting diodes), but in the 1960s their electrical properties were not well understood.[13]

One of the few people at RCA interested in solving this mystery was Sol Harrison, a physicist who carpooled with Heilmeier to Princeton. Harrison firmly believed that organic materials would become valuable additions to the circuit designer's tool kit, and he soon persuaded the younger engineer to join his research group in the DSRC's Electronic Research Laboratory (ERL).[14] One year later, Heilmeier submitted a PhD thesis on phthalocyanine in which he predicted that one day organic semiconductors would offer manufacturers "more flexible and cheaper materials, lower requirements for purity and simpler fabrication."[15]

Harrison and Heilmeier's organic semiconductor experiments were part of an industry-wide push to miniaturize electronic circuitry that was being driven by military demands for lighter, more reliable equipment.[16] RCA's first step in this direction was the micromodule, a self-contained array of uniformly shaped components that were layered on ceramic wafers and encased in a cube of epoxy resin. DSRC researchers also considered other approaches, including integrated semiconductor circuits and thin-film, cadmium sulfide transistors deposited on glass, which had been developed by RCA physicist Paul Weimer.[17] Compared with these alternatives, Harrison's research, which aimed to eliminate the need for individual circuit components by tailoring materials' chemical structures to perform complex electrical functions, was only in its beginning stages. Nonetheless, this "molecular electronics" program attracted attention from the US Air Force, which awarded his group a contract "to explore the possible uses of organic materials in solid-state devices."[18]

The air force assigned Harrison's group several molecular electronics projects, and Heilmeier found two especially compelling: organic thin-film transistors (as opposed to Weimer's inorganic devices) and electro-optic light modulators.[19] He divided his attention between both topics until the second half of 1963, when he postponed his transistor research to focus on light modulation.[20] He admitted later that this decision was equal parts technical and personal. "The laser had emerged on the scene at about this time and was commanding most of the attention at the Laboratory. I wanted some of the action."[21]

DSRC researchers had long recognized the laser's value in communication systems, particularly if high-speed modulators could be developed to enable the use of optical or infrared light to transmit information.[22] Unfortunately, most modulators—crystals whose refractive index changed when subjected to an electric field—were difficult to synthesize or suffered from prohibitive power requirements.[23] Inspired by his research into organic compounds such as phthalocyanine, which were simpler to grow and manipulate, Heilmeier "began asking the question . . . whether there were any molecular crystals that might make good optical modulators."[24]

He decided to concentrate on hexamine, a material whose crystalline structure was similar to existing modulators.[25] By November 1963, he had confirmed that hexamine's refractive index could be altered electronically and derived a mathematical relationship between the optical properties of a hexamine crystal and the behavior of electrons in its component molecules.[26] At the same time, he concluded that hexamine modulators would be infeasible because large ingots of the material were soft and tended to fracture easily.[27]

Rather than abandon hexamine entirely, Heilmeier proposed creating a new kind of modulator. If hexamine's bulk properties reflected those of its individual molecules, he reasoned, then it should be possible to mimic the properties of a large crystal by embedding

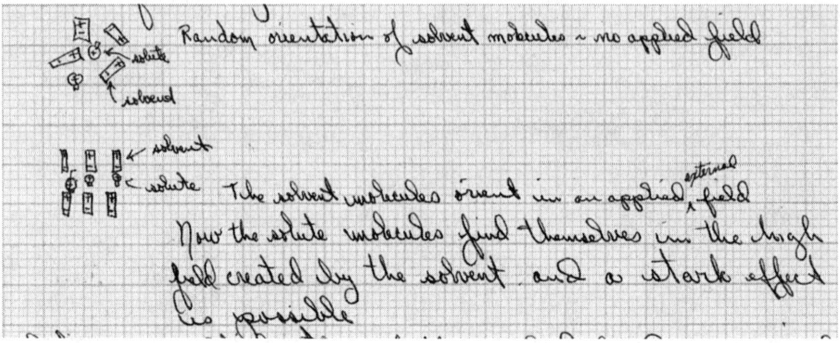

Figure 3.1. In this June 1964 notebook sketch, George Heilmeier imagined how a dye dissolved in a polar solvent might electronically modulate the passage of laser light. (Heilmeier, Notebook 24160, 72. David Sarnoff Library Collection, courtesy of Hagley Museum and Library.)

a batch of “guest” hexamine molecules in a clear plastic “host.” This “noncrystalline” modulator would theoretically possess identical properties to a sample of pure hexamine without any of the handling issues associated with comparably sized single crystals.[28]

Throughout early 1964, Heilmeier struggled to transform this concept into a functioning prototype, but hexamine’s softness and solubility in water made it tricky to encapsulate.[29] When the desired electro-optic effect failed to manifest in any of his test samples, he decided to forsake solid modulators altogether. Realizing that liquid solutions could affect polarized light in a similar fashion to crystalline solids, in June he sketched out a new type of modulator with an unnamed solute mixed into a highly polar solvent (fig. 3.1).[30] Before an electric field was applied, the solute molecules were randomly distributed among the elongated polar molecules of the solvent. The lack of internal organization meant that light would pass through the solution unaffected. When subjected to an external voltage, the solvent molecules would align themselves in rows parallel to the resulting field. “Now the solute molecules find themselves in the high field created by the solvent,” Heilmeier concluded, suggesting that these conditions would alter the absorption spectrum of the solute because of a localized Stark effect—the shifting and splitting of spectral lines under an electric field.[31]

While Heilmeier initially had a "host" solvent in mind for these new liquid laser modulation experiments—nitrobenzene—he did not specify what material would serve as his solute. A typed research proposal dated July 27, 1964, referred only to a "guest dye." In this document, Heilmeier also hinted for the first time at a possible display application for this research:

> The absorption spectrum (or color) of a dye in solution varies widely depending on the solvent. This effect can in some cases be attributed to the influence of the local high fields of the solvent molecules on the solute dye. We are interested in investigating the possibility of producing such effects by use of an external electric field. This would enable one to produce color changes that would be a function of an electrical signal. . . . Such a system would be useful for light modulation and displays.[32]

In the end, Heilmeier decided against the use of nitrobenzene as his solvent when his calculations indicated that in "the smaller external fields which are practical to apply" the resulting molecular realignment would not induce a spectral shift.[33] Luckily, while seeking an alternative solvent better suited to the task, he came across a pair of articles published the previous year by another DSRC researcher. His name was Richard Williams.

FROM LIGHT VALVES TO LIQUID CRYSTALS

The path that led Richard Williams to Princeton was hardly straightforward. After obtaining a doctorate in physical chemistry from Harvard in 1954, he received a job offer from Eastman Kodak, but before he could accept the position, fate—in the form of the US military—intervened. Despite having served in the navy at the end of World War II, the army drafted Williams and sent him to basic training at Fort Knox. Soon thereafter, he was reassigned to a chemical research facility in Edgewood, Maryland. Presumably, he noted in an oral history interview, the move resulted from his superiors' realization that "I had a Ph.D., which gave me a little bit more résumé than the usual hillbilly from Kentucky with whom I had been spending a couple of months."[34]

Just as Williams's degree facilitated his transition from boot camp to the laboratory, his academic connections helped him secure employment at the conclusion of his military service in 1955. His dissertation advisor—and future presidential science advisor—George Kistiakowsky, needed a new instructor to supervise Harvard's physical chemistry laboratory, a post that offered opportunities for independent research and that might serve as a springboard to a faculty appointment. Williams took the job, but after three years he concluded that a professorship was not in the cards. Instead, his gaze turned to industry, and in the summer of 1958, he became a member of the DSRC's technical staff.[35]

Upon arrival at RCA, Williams joined a materials science group responsible for the analysis of phosphors and photoconductors. Along with photoconductivity specialist Richard Bube, he launched a long-term project to determine the electrochemical properties of cadmium sulfide.[36] This research led to a solo study of the effects of strong electric fields on the absorption spectra of crystals. "I did some experiments. Some with cadmium sulfide and with some other materials. Lead iodide was one of them," he explained. "Colored materials, where if you had the right wavelength of light and you applied a voltage to it, you could shut off the light absorption near the absorption edge. It would shift the absorption edge so, in effect, it went black."[37]

Even as the DSRC pivoted away from fundamental research following the 1958 consent decree, Williams was able to continue his light modulation experiments.[38] He had no consumer application in mind for this work, although he briefly speculated on the possibility of incorporating a high-speed light shutter into pilots' goggles to protect their eyes from the flash of an atomic bomb.[39] Still, the DSRC was America's leading television research facility, so perhaps it was inevitable that by early 1962 Williams started wondering about "slow light valves" that operated at frequencies compatible with RCA's color broadcasting standard rather than the high-frequency switching induced in lead iodide.[40]

Williams's notebooks contain no references to RCA's previous mural television projects, but as he later commented, "everybody knew that if you had a flat display that would be a good thing. . . . It was so obvious I would think it wasn't even discussed."[41] Early on, however, he did not envision building anything as complicated as a TV display. Without integrated circuits, he noted, "you could conceive of making a pattern of those [light valves], and that was easy. And beyond that, it was hard to think very hard [or] think very much about it."[42]

Williams's ruminations on the possibility of creating a slow light valve eventually led him to consider the electro-optic properties of liquid crystals. These materials had been an object of scientific study since the late nineteenth century, when Austrian botanist Friedrich Reinitzer discovered a compound (cholesteryl benzoate) that did not undergo a simple transition from solid to liquid upon heating. Instead the material possessed two distinct melting points. At the first (146°C), it changed from a solid into a cloudy liquid, and at the second (179°C) it became transparent. Reinitzer forwarded this information to Otto Lehmann, a physical chemist who determined that in its cloudy intermediate state—later referred to as the "mesomorphic state" or "mesophase"—the compound retained its fluidity but refracted light like a solid crystal, indicating a degree of internal organization greater than most liquids. Lehmann described these substances as *Fliessende Krystalle* (flowing crystals), and within twenty years of Reinitzer's initial observation, European scientists reported over two hundred compounds with similar properties.[43]

The growing number of mesomorphic materials prompted French crystallographer Georges Friedel to propose a classification scheme based on their molecular orientation.[44] Most liquid crystals possessed elongated, cigar-shaped molecules. If these cigars were lined up with their long axes parallel to one another and packed into tight layers, the compound took on a thick viscous appearance, leading Friedel to refer to them as *smectic*, a Greek word meaning

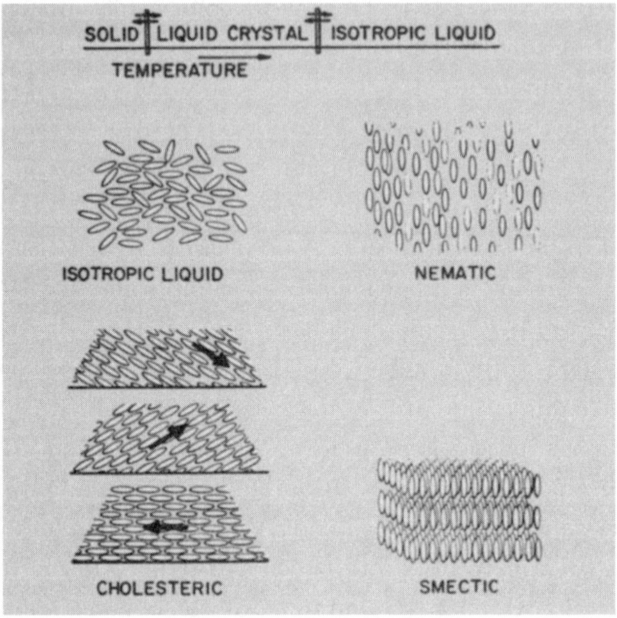

Figure 3.2. Scientists classify liquid crystals into three categories based on their molecular organization. The majority of LCD work at RCA involved nematic liquid crystals. (G. H. Heilmeier, "Liquid crystals: the first electronic method for controlling the reflection of light," *RCA Engineer* 15, no. 1 [June–July 1969]: 15. David Sarnoff Library Collection, courtesy of Hagley Museum and Library. Scan courtesy of Linda Hall Library of Science, Engineering & Technology.)

"soap-like." If the molecules could slide past each other while keeping their long axes parallel, like matches in a box, they were called *nematic*, another Greek term, referring to their "threadlike" texture when viewed between crossed polarizers. The third group of crystals traced its name to the cholesterol esters in which they were first observed. These "cholesteric" compounds had molecules arranged in layers like a nematic crystal, which were then stacked on top of each other in a helical pattern that scattered light from its surface. Changes in temperature could change the dimensions of this spiral, altering its absorption properties and, consequently, its color (fig. 3.2). Heating any of these materials beyond a certain point caused their molecules to become increasingly disorganized, so they behaved like an ordinary (or *isotropic*) liquid.[45]

By 1962, the novelty of discovery had dwindled for liquid crystals. Once the research province of such luminaries as Max Born, Louis de Broglie, and J. D. Bernal, they now merited only a cursory mention in physical chemistry textbooks.[46] With the exception of a handful of color-changing temperature sensors using cholesteric crystals—an idea pioneered by Westinghouse physicist James Fergason and later popularized through the mood ring fad of the 1970s—industrial applications were almost nonexistent.[47] Williams had learned about liquid crystals in graduate school, but he had never worked with them in a laboratory setting. Nevertheless, after surveying the literature on them in the DSRC technical library, he felt confident enough to proceed with his experiments. "I was quite aware of the general nature of them, and some of them were commercially available in catalogs of organic compounds. And so that seemed like a possibility," he recalled. "So that's where I started."[48]

ENTERING A NEW DOMAIN

The apparatus that Williams assembled for his April 1962 exploration of "low frequency light modulation using liquid crystals" resembled his earlier lead iodide experiments.[49] In both instances, he placed a crystal sample between two glass plates whose inner faces were lined with a transparent conductive coating that allowed him to control the electric field applied to the sample. Both times, he placed the cell on a microscope stage that was illuminated from below in order to observe any changes in light transmission. The main difference in this new setup was the inclusion of a heating element to keep the liquid crystals' temperature between its two melting points. In the case of para-azoxyanisole (PAA), the readily available mesomorphic compound that was the subject of his first trials, this meant warming the sample to anywhere from 116°C and 134°C.

Once the sample was properly heated, Williams connected it to a power supply and switched on the voltage. At first nothing happened. Light passed through the cell unimpeded, even as he increased the strength of the field. Suddenly, when the field's magnitude reached 1,000 V/cm, Williams observed a curious "crinkling effect" as long parallel regions appeared in the previously transparent liquid crystals. Raising the field further, to 10,000 V/cm, caused these regions to disappear as a "vigorous turbulence" spread throughout the sample. Once the field was turned off, the PAA reverted to its settled state.[50]

Williams repeated the experiment several times, applying both AC and DC voltages and viewing the sample with polarized light—a common technique used to reveal otherwise invisible structural details of crystalline materials. In every case, he saw that at a fixed threshold voltage "a stable pattern is formed which becomes turbulent as the voltage increases."[51] This pattern, which Williams referred to as a "domain structure," also decreased the amount of light passing through the sample. "You could just see that without any instrumentation," he remembered. "You apply the voltage, and the liquid begins to look cloudier than it did. And that's because the light-scattering properties are changed. Quite significantly, so it's obvious. It doesn't take hours of staring in a dark room or anything to see it. It was easy" (fig. 3.3).[52]

Williams wasted little time replicating this experiment for other members of his research group, including his supervisor, Simon Larach. Larach, "who was very interested in forwarding the concept," urged him to prioritize his liquid crystal research and compose a patent disclosure based on his findings.[53] Over the coming months, Williams followed Larach's advice, conducting detailed studies of PAA switching at different temperatures and observing the formation of domains—later named "Williams domains" in his honor—in other liquid crystals.[54]

In November 1962, he also submitted a patent for "an improved electro-optical element" utilizing nematic compounds.[55] Williams was not the first person to suggest a display device based on liquid crystals. That honor belonged to a pair of British employees at the Marconi Wireless Telegraph Company who in 1934 described how

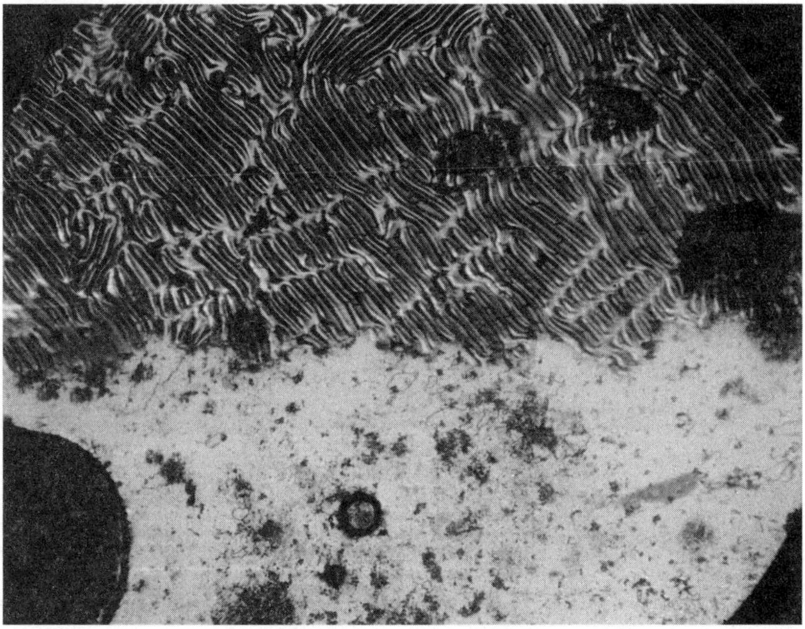

Figure 3.3. Richard Williams took this photograph of the domain structure of p-azoxyanisole (PAA) on April 13, 1962, three days after his first liquid crystal experiments. He placed the sample between crossed polarizers and viewed it through a microscope under 90× magnification. (Williams, Notebook 15811, 78. David Sarnoff Library Collection, courtesy of Hagley Museum and Library.)

mesomorphic materials could serve as a shutter in electromechanical television or facsimile systems.[56] Their proposal was similar to Williams's original concept of a low-frequency light valve, but now he imagined a more ambitious setup. Referring directly to Jan Rajchman's flat-panel prototype, he alluded to the possibility of constructing a mural TV that displayed images on an array of liquid crystal cells. While "a photograph or picture can be displayed on a device of this type," Williams pointed out, the switching speeds of his liquid crystal samples were also "compatible with present television standards."[57]

Considering these optimistic claims, one might expect Williams to have led the charge to build prototype liquid crystal displays, but his laboratory notebooks suggest the opposite. After filing his patent, he concluded that the wonderful applications it described would not be forthcoming anytime soon because of the absence of room-temperature mesomorphic compounds. In addition, Williams found "it was hard to get people interested in the device end because the typical electrical engineer who was working had never heard of liquid crystals. This is a nutty thing."[58]

While the obscurity and physical constraints associated with liquid crystals hampered efforts to build displays based on Williams's discoveries, so too did the absence of a strong advocate for the new technology. For his part, Williams viewed his work on liquid crystals as one project among many and balanced attempts to decipher the mechanism of domain formation alongside electro-optic investigations of solid, inorganic crystals such as cadmium sulfide and sodium chloride.[59] He was also willing to leave the task of display construction to others, most notably Simon Larach, who built a demonstration model with physicist Ross Shrader in May 1962 using Williams's "optical-field-effect."[60] The device resembled Williams's original liquid crystal cell except the two plates were each lined with four strips of conductive coating, which he arranged at right angles to form a grid. Once the sample was heated, Larach could cause domains to form at any given intersection by applying voltage to the corresponding set of electrodes. This "x-y array" echoed the display concept outlined in Williams's patent, which is not surprising because the chemist would have consulted with Larach while formulating his ideas.[61]

In any case, this work led Larach—not Williams—to become the most prominent early spokesman for LCD development within RCA. He and Shrader showed off their new prototype to DSRC head James Hillier as well as other researchers with an interest in new display technologies, including SRL personnel such as Allen Barco and David Kleitman and Computer Research Laboratory (CRL) head and flat-panel proponent Jan Rajchman. Williams was often in attendance, but he preferred to let others do the talking. He proved so effective at fading into the background during these demonstrations that CRL engineer Bernard Lechner later admitted "I had no idea at the time that Dick Williams was involved at all."[62]

Larach's enthusiasm, however, could only take Williams's ideas so far. His responsibilities as group head left him ill suited to carry the liquid crystal project forward. Nor would he compel Williams to return to the subject after the latter decided to participate in a military-sponsored study of ionic solids under high electric fields.[63] Williams published two articles on domain formation in July 1963, but a month later he departed for a nine-month assignment at RCA's Zurich laboratory, leaving any discussions of liquid crystal displays to fade on the other side of the Atlantic until George Heilmeier came along.[64]

PLEOCHROIC POSSIBILITIES

Heilmeier had met Williams soon after arriving in Princeton, and their shared interest in electro-optic materials provoked frequent conversations. "We talked all the time," Williams recalled. "We often had lunch together, saw each other, walked into the other's lab, and so on."[65] Heilmeier, in turn, referred to Williams as "one of the finest physical chemists I have ever met. He just had a wonderful physical feel for things."[66] In multiple journal articles, Heilmeier also thanked Williams for their "helpful discussions."[67]

There is little documentary evidence showing how these interactions with Williams influenced Heilmeier's laser modulation experiments. The two men collaborated on a study of the ferroelectric properties of liquid crystals, but their laboratory notebooks contain no references to any discussion of Heilmeier's July 1964 proposal "to investigate the spectra of dyes in a liquid crystal medium."[68] However, in a 1998 oral history, Heilmeier acknowledged the importance of Williams's domain experiments to his thinking. As it became clear hexamine would not serve as an effective light modulator, he explained, "I looked at what Dick Williams was doing and I thought about the following: I thought, suppose I could use an electric field to align a liquid crystal? Which was what Dick was doing."[69]

Though Heilmeier probably made the move from polar solvents to liquid crystals independently, he did not do so alone. Unlike Williams, who conducted his PAA experiments by himself, Heilmeier's liquid crystal work depended, even at this early stage, on the contributions of numerous individuals. The most noteworthy of these was Louis Zanoni, a naval veteran from Trenton who graduated with honors from RCA's technical school before starting at the DSRC in 1957. Zanoni established himself as a skilled technician with a particular aptitude for requisitioning materials and equipment. "I was not a cleanup guy. I was a technician, and I had lots of contact with all of the services of the laboratory," he explained. "I could go to any shop, any place, [and] ask them for something. They would help me, provide anything I needed."[70]

Zanoni had impressed Heilmeier with his "extremely capable technical assistance" during the hexamine project, and the two formed a close partnership that lasted for the remainder of the decade.[71] Two months after Heilmeier expressed interest in studying dyes dissolved in liquid crystals, he and Zanoni commenced their project by reproducing Williams's PAA experiments. After confirming the existence of domains, they mixed various dyes into the liquid crystal and repeated the procedure. If Heilmeier's predictions were correct, the field-aligned liquid crystal molecules would induce a change in the dye's absorption spectrum, leading to a color shift. Alternatively, the domain pattern could be used to physically manipulate a pleochroic dye, whose color depended on its molecular orientation with respect to polarized light.[72]

Results during the first two weeks proved frustratingly inconclusive. Heilmeier worried that the interior surfaces of the glass plates had affected the initial orientation of the liquid crystal molecules. He also questioned the selection of PAA for these experiments, since its yellow color could mask a possible color change.[73] To resolve the first issue, Heilmeier and Zanoni created a thorough cleaning regimen for their glass substrates: washing slides in strong acids, rinsing them with distilled water, and rubbing them dry in a single direction to ensure uniform molecular alignment.[74] For improved materials, Heilmeier approached Joel Goldmacher. Goldmacher was the first organic chemist to join the staff of the DSRC, having been hired in 1963 to support Sol Harrison's molecular electronics initiative.[75] He had synthesized hexamine crystals for Heilmeier's earlier experiments and later supplied the engineer with the names of PAA-soluble dyes. Now he partnered with Zanoni to synthesize and test new samples of purified liquid crystals. Heilmeier also received assistance from Daniel Ross, another chemist with an interest in optical modulators who offered him additional dyes for evaluation.[76]

The demonstration of Heilmeier's predicted optical effect in late October 1964 was therefore a collaborative enterprise. The experiment revolved around a dye (supplied by Ross) dissolved in a liquid crystal (courtesy of Goldmacher) and secured between two transparent electrodes (prepared by Zanoni). In place of the yellowish PAA, Heilmeier substituted a colorless nematic compound called MOCA (short for trans-p-methoxycinnamic acid), which turned pinkish red when combined with Ross' dye. He and Zanoni then fabricated a cell, placed it on a heated microscope stage, and viewed it with polarized light. The procedure was nearly identical to Williams's, but the outcomes differed greatly. As Heilmeier recorded in his notebook, when 50 volts was applied to the sample, "the region in the field became almost colorless and returned to red when the field was removed."[77]

These results confirmed that it was possible to alter the absorption spectrum of "guest" dyes dissolved in a liquid crystal "host." Additional testing revealed that the effect was not limited to MOCA but to a variety of dyes and liquid crystals that were better able to withstand long-term exposure to high temperatures. Heilmeier remained somewhat uncertain whether the color shift was caused by electric fields acting on the dye molecules, as he originally imagined, or their physical reorientation relative to polarized light. Further experiments would lead him to endorse the latter explanation in the spring of 1965, but long before then, the engineer had started to consider commercial uses for liquid crystals.[78]

Immediately after recording the results of his first MOCA experiment, Heilmeier compiled a list of applications for the guest-host effect with an eye toward a future patent disclosure. Some of these, like "an electrically controlled light shutter for photographic applications or laser Q switching" could be traced to his earlier interest in

Figure 3.4. Through selective application of conductive coatings, George Heilmeier and Louis Zanoni used guest-host color switching to generate simple patterns such as the RCA "meatball." (Zanoni, Notebook 16971, 93. David Sarnoff Library Collection, courtesy of Hagley Museum and Library.)

electro-optic modulators. Others, such as a spectrophotometer to analyze pleochroic dyes, would only be of interest to industrial chemists. But for Heilmeier, the most enticing use for his color switching liquid crystals was in display panels. "Since large area, flat panels can be fabricated & driven by low voltages," he wrote, "this device should find application in displays where such properties are desireable."[79]

Where Williams's patent had presented liquid crystal televisions as a long-term prospect at best, Heilmeier was confident that the guest-host effect had brought them within arm's reach. He and Zanoni were already able to produce simple images, like RCA's circular logo—nicknamed "the meatball"—by etching patterns into their slides' conductive coatings (fig. 3.4).[80] In Heilmeier's mind, the leap to moving pictures was only a matter of time and resources, and he was confident he could obtain both. "It was Fall 1964," he would later reminisce. "The wall-sized flat panel color TV was just around the corner—all that you had to do was ask us!"[81]

MOVING TO THE REFLECTIVE MODE

For all of his newfound fascination with flat-panel displays, Heilmeier's guest-host investigations had been carried out under the auspices of RCA's quantum electronics group. His supervisor, Henry Lewis, supported this research and arranged demonstrations of electronic color switching for ERL director William Webster and other high-ranking DSRC staff.[82] But the more Heilmeier's interests drifted from laser modulation to display systems, the less relevant they seemed to the ERL's stated focus on "integrated electronics, semiconductor devices, energy conversion, and quantum electronics (lasers and masers)."[83] If Heilmeier wished to create liquid crystal televisions, he would need assistance from the SRL, which was officially responsible for developing new display devices.

Eventually he found an ally in David Kleitman. More than any other SRL group leader, Kleitman had demonstrated an openness to harnessing new electro-optic phenomena for use in electronic displays. Though his personal eccentricities, such as tinkering with a collection of old cars on his front lawn, led some at the DSRC to refer to him as "Mad Man" Kleitman, the members of his team, like engineer John van Raalte, respected his management style.[84] "We didn't have a whole lot of philosophical discussions," Van Raalte observed, "but he had a broad interest in technology, and I think at the time he was willing to let people search for themselves and pursue thoughts, ideas, [and] interesting avenues."[85]

It was in all likelihood this receptiveness to new concepts that prompted Heilmeier to cross organizational lines and approach Kleitman about the possibility of a joint ERL/SRL project to consider possible display applications of liquid crystals. Kleitman responded by arranging for Lucian Barton and Warren Moles to watch Heilmeier demonstrate the guest-host effect and assist with the characterization of new mesomorphic compounds. Barton would consider strategies to broaden the temperature range in which these materials existed in the nematic phase, while Moles examined their switching properties.[86] Heilmeier consulted with both men while continuing his electro-optical studies, but formal coordination between ERL and SRL was limited to occasional meetings in Kleitman's office. At times, this lack of organization could be disheartening. "Needed at this time—discussion of the various aspects of the problem," Warren Moles wrote in a March 1965 notebook entry. "It may be that we should get together the various people concerned with the project and attempt to define more clearly what we are after and how we will get there."[87]

The transition from a loose agglomeration of researchers to a DSRC-sanctioned liquid crystal group with well-established objectives coincided with Heilmeier's growing recognition of the guest-host effect's shortcomings. With the benefit of hindsight, its flaws were obvious:

> The dyes and their liquid crystal hosts were not stable over long periods of time in applied fields, the effect was sensitive to surface orientation effects, it required polarized light, it was viewed in transmission, it required heating to maintain the host in its nematic phase, uniformity was a problem, and so on.[88]

Until the end of 1964, Heilmeier had been confident that new materials and savvy engineering could overcome these obstacles. As before, he relied on Joel Goldmacher for the former and Louis Zanoni for the latter, but the guest-host effect's problems appeared intractable. In the first quarter of 1965, Goldmacher synthesized more than half a dozen new liquid crystal compounds. None of them offered a comprehensive solution to the litany of issues confronting electronic color switching, but in late February, one of them spurred Heilmeier to revisit a phenomenon Williams had observed three years earlier.[89]

The material in question was called DHOBABB.[90] Like the PAA that Williams and Heilmeier used in their earliest liquid crystal experiments, it was a pale yellow compound that formed domains when subjected to an electric field. If the field was increased by another order of magnitude to ~20,000 V/cm, the sample underwent what Heilmeier described as either "bubbling" or "stirring."[91] There was a precedent for such turbulence. Heilmeier had observed it in other liquid crystals, and Williams's journal articles described how a "gentle stirring" in his PAA samples had evolved into a "vigorous agitation" as he increased the applied field.[92] What captured Heilmeier's attention in this instance was a noticeable increase in DHOBABB's contrast once the turbulence appeared. "The contrast ratio for field on & off in reflected light (no polarizers) was 6.3:1," he wrote, indicating on the next page of his notebook that "a standard TV monitor in our TV studio yielded a contrast ratio which was only 4:1."[93]

A few days later, Heilmeier ran the experiment again with a slightly different setup. "Previously we had noted the high contrast ratio obtained in reflection for the DHOBABB system when fields high enough to cause stirring were used," he wrote in his notebook. "It was of interest to check what resolution could be obtained for such a system in view of the field induced cavitation."[94] To that end, he asked Zanoni to etch the transparent conductive coating on one piece of glass so that it formed the RCA meatball and activate the field. The results were disappointing. "The pattern is barely visible," Heilmeier lamented before abandoning the idea of an LCD based on this stirring effect, seemingly for good.[95]

Throughout the spring, Heilmeier evaluated Goldmacher's new crystals, paying close attention to their ability to modulate the passage of polarized light. During one such test, with a compound called APAPA (anisylidene p-aminophenyl acetate), he encountered a logistical difficulty. "It has been noted that not all APAPA samples orient such that they are dark between crossed polarizers & light when field is applied," he wrote in May 1965.[96] For some reason, the material was sticking to the interior surface of his glass slides, preventing its molecules from moving when he applied an electric field. After a series of acid washes failed to resolve the problem, Heilmeier recalled the stirring effect he had seen in DHOBABB and decided to adapt it as a cleaning technique. After doping the APAPA with a small amount of another liquid crystal called BBA, he subjected the sample "to violent stirring (by field) to produce mixing."[97] Not only did the mixture exhibit uniform switching behavior, but the procedure had an unexpected side benefit. As Heilmeier reported,

> Some time ago we noted the high reflectivity of the stirred state of some nematic systems relative to the quiescent state without field. At that time the observations were made on yellow materials which responded quite slowly. We have now reviewed this effect in APAPA + ½% BBA and found it to be quite striking.[98]

At long last, Heilmeier had uncovered an electro-optic effect that might enable assembly of a high-contrast LCD whose brightness would not suffer from reliance on filters to polarize light. As he had with electronic color switching, he invited DSRC personnel to watch as transparent APAPA cells turned milky-white thanks to what he initially termed "the reflective or scattering mode" before settling upon "dynamic scattering."[99] Some of these visitors, like Bernard Lechner, recalled this behavior from Larach's demonstration of Williams's research three years earlier.[100] But others, including David Kleitman, believed the reflective mode was the key to devising the long sought-after replacement for the cathode-ray tube. As John van Raalte, who would join Warren Moles to work on LCD addressing systems put it, "Kleitman was the one who, as I recall, recognized the importance of a cell going from clear to white or black to white and said, 'That's fantastic! We've got to start a secret project on this.'"[101]

SECRECY BETWEEN RED COVERS

At the outset, the liquid crystal group consisted of only seven people in addition to Kleitman, who secured formal authorization in the spring of 1965 for a joint SRL/ERL study of liquid crystal display applications and who served as project coordinator.[102] It is a testament to Heilmeier and Kleitman's persuasive abilities that despite this small size, the DSRC's management declared its work "company private," a status previously granted to only a handful of products with potentially significant ramifications on RCA's existing product lines.[103] In daily practice, this secrecy was nominal. Information flowed freely within the group, and nonaffiliated DSRC personnel, such as Williams, could still consult with its members.

The major consequence of receiving company-private status was the prohibition of any discussion of LCDs outside the DSRC until the company guaranteed its legal claims over the technology.[104] RCA did not cut itself off completely from the scientific community. Heilmeier and Williams received permission to travel to Kent State University in August 1965 for the first International Liquid Crystal Conference with the understanding that references to displays should be omitted from their presentations.[105] This blackout also extended to journal articles that RCA staff published on the chemistry and physics of nematic materials as well as the technical reports or engineering memorandums that researchers normally submitted to the DSRC's library.[106] The only sanctioned outlet for extended written discussions of LCDs were the confidential progress reports that Kleitman compiled based on research digests from the rest of his team. Company administrators carefully monitored the circulation of these documents, which were distinguished by their bright red covers, granting access only to liquid crystal researchers and DSRC management.[107]

Kleitman oversaw the composition of these reports through June 1966. At that point, he abdicated responsibility to Heilmeier, who having served as de facto group head now earned an official promotion to that rank. By the end of the following year, clashes with management over the development of a television-based facsimile system known as Homefax prompted Kleitman to depart RCA to become director of R & D at Signetics—a prominent Bay Area integrated circuit manufacturer.[108]

The liquid crystal research team would more than double in size over the next three years. Its members included chemists, physicists, and engineers drawn from the ERL and SRL as well as a sizable delegation from the CRL recruited because of their familiarity with flat-panel addressing circuitry. Heilmeier was the driving force behind this expansion and emerged as an effective leader of an increasingly complex project. He smoothed over institutional and disciplinary divisions within the group by encouraging an atmosphere of playful camaraderie. In between experiments, members of RCA's liquid crystal team played touch football in the DSRC courtyard, caught fish in the nearby Millstone River for a makeshift aquarium, and ordered pastel lab coats from *Playboy* magazine.[109]

One thing that did not change under Heilmeier's leadership was the compilation of the red-covered secret reports summarizing the liquid crystal group's findings. Their publication rate varied, but Heilmeier retained the organizational structure that Kleitman had set in 1965. Each began with a materials update from the chemists, moved on to theoretical analyses of the physical mechanisms underlying dynamic scattering, and concluded with discussions of device fabrication. Taken as a whole, these reports provide a continuous digest of the liquid crystal group's activities and evidence of Heilmeier's ability to resolve complicated technical problems by himself and in coordination with others.

SEARCHING FOR AN IDEAL MIXTURE

Demonstrating guest-host color switching and the reflective mode may have been sufficient to establish a research group, but the LCD project's long-term success was contingent on Heilmeier's ability to procure improved materials. The idea of a liquid crystal television display was laughable so long as the chemicals involved only existed in a mesomorphic state within a narrow range of high temperatures. The APAPA used in Heilmeier's latest experiments, for example, had to be heated to between 83°C and 110°C—rendering its inclusion in consumer products unsafe and impractical. Heilmeier had already broached the need for low-temperature nematic compounds with Joel Goldmacher and Lucian Barton, but neither was able to resolve the issue.[110] In truth, the group's chemists still lacked a clear grasp of the material logic of liquid crystals.[111] With the formation of a more structured research program, they would gain additional reinforcements as they grappled with the mysteries of mesomorphism.

The chemistry cohort of the DSRC's liquid crystal group was small, but it embodied a broad range of professional experience. Goldmacher, the Brooklyn native responsible for introducing organic chemistry to the DSRC, worked with Lucian Barton, a veteran of the Polish and British armies who trained as a chemical engineer in Turin before immigrating to the United States.[112] Joining them was Jean Kane, the only woman in the LCD group's original roster, who Kleitman hired shortly after she had earned a master's degree in chemistry from Mount Holyoke College. Kane assisted Goldmacher with the synthesis of new organic compounds until June 1965, when she withdrew on maternity leave.[113]

Stepping in to take her place was Joseph Castellano, a classmate of Goldmacher's from the City College of New York. Castellano had previously held a position testing rocket propellants at the Reaction Motors Division (RMD) of Thiokol Chemical. Although he enjoyed the technical aspects of his work, like many scientists and engineers during the 1960s, he became disillusioned by RMD's close ties with the military and the destruction that the company's missiles had inflicted in Vietnam.[114] These misgivings and uncertainty about the renewal of his project's funding led him to ask Goldmacher if there were any positions at RCA. "As luck would have it, there was indeed an opening for an organic chemist to work in the newly emerging field of liquid crystals," Castellano wrote later, "I was certainly interested, but I still had no idea what liquid crystals were all about."[115]

Fortunately, Castellano was a quick study, and by the beginning of July, he, Barton, and Goldmacher had agreed on a three-part strategy to create a room-temperature liquid crystal. "Encouraging results with APAPA" convinced all three to focus on compounds with a similar two-ring structure.[116] Goldmacher, who had the most experience synthesizing mesomorphic compounds, would seek out new materials and purification techniques, while Barton and Castellano modified APAPA in two distinct ways in hopes of expanding their nematic range. As Barton doped liquid crystals with small amounts of structurally similar chemicals, Castellano directly altered APAPA's molecular structure to weaken the intermolecular forces within a sample and lower its melting point.[117]

Neither Barton nor Castellano's approach was immediately successful, but by March 1966, Castellano was able to generate a single APAPA derivative—later known as APAPA-12—with a melting point of 50°C.[118] This result was still a long way from the 25°C required for room-temperature operation, but during the course of this work Castellano uncovered a new property of his liquid crystal samples with promising implications. In October 1965, he showed that combining equal amounts of two liquid crystals could produce a mixture whose melting point was actually lower than either of its constituents. In one instance, he mixed two APAPA derivatives, both of which had approximate nematic ranges of 82°C –110°C, and the combination's melting point plummeted to 47°C. Since the mixture's upper transition point was almost unchanged, Castellano had, in one fell swoop, doubled the liquid crystal's operational temperature range.[119]

As it turned out, this effect was not limited to binary mixtures. Adding a third APAPA derivative lowered the mixture's melting point to 40°C. Heilmeier praised this new ternary mixture in the group's progress reports, noting that the chemists' work "enables us to study such important device parameters as speed of response, contrast ratio and resistivity as a function of temperature."[120] Goldmacher and Barton stopped searching for individual room-temperature compounds and concentrated on supplying Castellano with more APAPA derivatives. Finally, in April 1966, they announced the creation of a ternary mixture with nematic range of 23°C–105°C.[121]

Having broken the room-temperature barrier once and for all, the chemists now focused on refining their procedures and exploring other properties of mesomorphic materials. Castellano, Goldmacher, and Barton accumulated more data about their liquid crystal mixtures and constructed three-dimensional phase diagrams that enabled them to calculate the proportions of various APAPA derivatives that would produce a mixture with a desired nematic range (fig. 3.5).[122] Meanwhile, Heilmeier and Jean Kane, who had returned to the DSRC on a

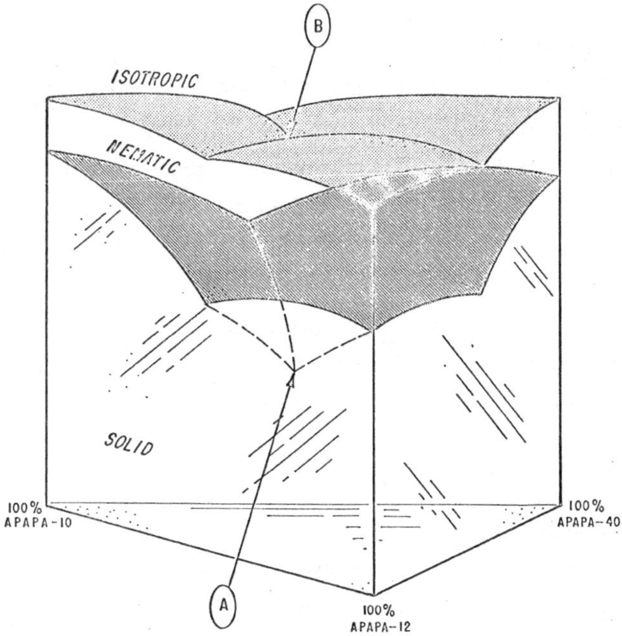

Figure 3.5. Three-dimensional phase diagrams like this one allowed chemists Joseph Castellano and Joel Goldmacher to deduce how different combinations of liquid crystals (such as the three APAPA derivatives shown here) would affect the mixture's nematic range. *A* and *B* indicate the eutectic points, the combinations that would lead to phase transitions (solid to nematic; nematic to isotropic) occurring at the lowest possible temperatures. (RCA Laboratories, *Progress Report: Liquid Crystals, Jan.–Mar. 1967*, 5. David Sarnoff Library Collection, courtesy of Hagley Museum and Library.)

part-time basis, began a series of life tests, building LCDs with electrodes made from different metals and observing how long they could operate before the liquid crystals decomposed. These studies con firmed that the presence of water or oxygen hastened the decay process, but determining why that occurred would have to wait until the group had a greater understanding of dynamic scattering itself.[123]

FIRING MOLECULAR BULLETS

Upon viewing the reflective mode for the first time in May 1965, Heilmeier had concluded that "the scattering of light by the domains when the material is stirring is responsible for the effect" but conceded in his group's first progress report that "the mechanism and cause of the stirring are unclear at this time."[124] Early on, he deemed such theoretical questions low priority compared with the chemists' material studies. "With all due modesty," he explained at the end of 1965, "we feel that the most we can hope for is a qualitative understanding of the field induced scattering from liquid crystals such as APAPA."[125]

But even a qualitative understanding would require some quantitative measurements, and by the spring of 1966, Heilmeier felt the time was right to begin collecting data related to the origins of dynamic scattering. "For some time, I have been thinking about the mechanism of stirring in APAPA type compounds," he wrote in a March notebook entry. "I feel that the stirring which produces the reflection properties should be viewed as a true turbulence and more attention paid to the hydrodynamic aspects of the phenomenon."[126] Viscosity measurements of the liquid crystals subsequently persuaded Heilmeier that the observed optical effects were not caused by "turbulence in the normal sense (i.e., inertial forces >> viscous forces)."[127] Something else was responsible, and in June he argued that "some insight might be gained by considering the wake of bodies moving through the medium"—specifically ions.[128]

As early as May 1965, Heilmeier had contemplated "the importance of ions in the behavior of nematic melts" and their involvement in domain formation.[129] More recent comparisons of purified liquid crystals implied that "ions present in the nematic phase play an important role in determining the electrical properties of the material."[130] In the summer of 1966, he resurrected the idea, suggesting that light scattering took place in the wakes formed by ions passing through a liquid crystal medium. To observe these disturbances, he and Zanoni designed a series of new cells that allowed them to view dynamic scattering from a variety of different angles.[131] These experiments strengthened his confidence in a new model of dynamic scattering.

According to this "momentum exchange" framework, applying

a strong enough electric field caused negative ions to travel across a liquid crystal cell, moving from the negative electrode (the cathode) to the positive one (anode) and colliding with neutral liquid crystal molecules along the way. These collisions induced "a turbulence (or actual flow) in the liquid crystal and its associated optical effect."[132] This model provided a somewhat satisfying explanation for dynamic scattering, but it was still incomplete. It was one thing to claim that ions were responsible for liquid crystals' electro-optic behavior, but Heilmeier could not explain the source of the ions or why some materials, like APAPA, were more conducive to dynamic scattering than others.

Heilmeier and Barton initiated a second set of investigations in early 1967 to answer these questions.[133] Previously, Heilmeier assumed that the ions responsible for dynamic scattering were ejected directly from the electrodes, but he and Barton now found that strong electric fields caused liquid crystal molecules to dissociate near the cathode.[134] Accelerated by the field, each of these ions acted like what Heilmeier later termed a "molecular bullet," careening through the sample.[135] At the same time, the electric field caused the cigar-shaped liquid crystal molecules to move into an organized pattern. In nematic compounds that exhibited dynamic scattering, the molecules lined up with their long axes parallel to the glass substrates and perpendicular to the field, forcing the ions to slam their way across the cell like boats through a log jam. The disruption of this structure led to turbulence that scattered light.[136]

As had been the case the previous year, this revised momentum exchange model left the LCD research group with several mysteries, such as the chemical reactions involved in ion formation. It also failed to account for several new electro-optic effects encountered during Heilmeier and Barton's investigations, such as the use of short current pulses to terminate dynamic scattering.[137] To accommodate broader research into these phenomena, Heilmeier enlisted new personnel into the research group. Physical chemist Alan Sussman studied the behavior of charge carriers in APAPA and partnered with Castellano to investigate the role of hydroxyl radicals in liquid crystal decomposition.[138] Another recruit was German physicist Wolfgang Helfrich, who constructed mathematical models to explain the formation of domains and their behavior under applied fields.[139] These new staff members enabled Heilmeier to scale back his theoretical research and concentrate on the liquid crystal group's primary goal: the construction of dynamic scattering displays.

FROM STATIC DISPLAYS TO THE ACTIVE MATRIX

Much as Heilmeier enjoyed playing the theorist, he identified himself first and foremost as an electrical engineer. Even before witnessing dynamic scattering in APAPA, he had wondered how an incoming television signal could be converted into an image on a liquid crystal screen. In February 1965, he composed a five-page notebook entry evaluating liquid crystals' compatibility with the conservative and radical addressing systems utilized in previous RCA flat-panel projects.[140] For further advice on how to proceed, he contacted two members of Kleitman's SRL group, John van Raalte and Warren Moles. The two concluded in August 1965 that "TV speed matrix addressing can be accomplished at present," but additional components would be required to isolate and store the signal at each of the 250,000 liquid crystal picture elements required to achieve full TV resolution. Existing diodes could fulfill this function, but they were not amenable to batch fabrication in integrated circuit arrays.[141]

Van Raalte and Moles anticipated either new materials or cell structures might simplify the construction of a matrix-addressed LCD. Until then, they decided to verify that liquid crystals could generate a high resolution static image using an "insulator addressed" system. Adapting a technique pioneered in the semiconductor industry, they coated a glass slide with a transparent conductor and a light-sensitive polymer called a photoresist. They then projected a photographic negative on the photoresist, hardening the areas exposed to light so they could not be removed with an acid rinse. When the liquid crystal was placed between the treated glass, the areas where the photoresist was dissolved (the dark areas of the negative) underwent dynamic scattering and appeared bright. Regions where the photoresist remained (the light areas of the negative) were insulated and stayed dark.[142]

The SRL researchers considered this new fabrication technique a proof of concept rather than a justification for further research and returned to studying matrix addressing in early 1966.[143] This did not prevent Heilmeier and Zanoni from appropriating their methods to construct displays featuring images of everything from TV test patterns to David Sarnoff.[144] Their displays operated at low power, since they did not generate any light of their own, and featured reflective backplates to redirect the large amount of forward-scattered light back toward the viewer's eye. A handheld "personalized flat display" confirmed this backplate did not impose any restrictions on indoor or outdoor viewing, although an adjustable lid helped to maximize contrast (fig. 3.6).[145]

Impressive as these static displays were, they did little to advance the group's long-term objective of a liquid crystal television display. In a midyear statement preceding the June 1966 progress report, Kleitman urged his colleagues to keep their eyes on the prize and redouble efforts to create both cathode-ray and matrix-addressed systems. "Even if the present x-y address emphasis is increased, and the project remains undistracted from this course," he cautioned, "some attention within the corporation should be given to the determination of the role and problems of an interim electron beam addressed product."[146]

The LCD group took Kleitman's recommendation to heart and before the end of the year could report progress along both fronts. In November 1966, John van Raalte successfully modified a CRT-based projection system to address a liquid crystal target. The display was only 2 inches square and mounted on the face of a bulky vacuum tube, but the pictures, taken from live televisions signals, could be thrown on to a much larger screen.[147] It was the first time an over-the-air broadcast appeared on an LCD, prompting Van Raalte to proudly declare in his notebook that the "World's first electron-beam-addressed, reflective television display was born" (fig. 3.7).[148]

Figure 3.6. Louis Zanoni holding a portable dynamic scattering display outside of the DSRC in spring 1966. (Courtesy of Louis Zanoni.)

Van Raalte's display marked the apex of the conservative approach to LCD addressing. His affiliation with the LCD group had waned since the summer of 1966, and he knew that his CRT setup "was clearly only practical for making large, projected images."[149] The smaller, direct-view displays Heilmeier had in mind would require matrix-addressing circuitry, so that fall he reached out to Bernard Lechner, who had just delivered his 1,200-element electroluminescent display to Wright-Patterson Air Force Base.[150] Lechner recalled discussing matrix addressing with Heilmeier before signing on to the LCD project. "I had conversations in those days with George [Heilmeier] and Lou [Zanoni] and Dave Kleitman and others—John van Raalte, I'm sure," he remembered, "about how we'd make a matrix display out of this when we get some materials that will work at room

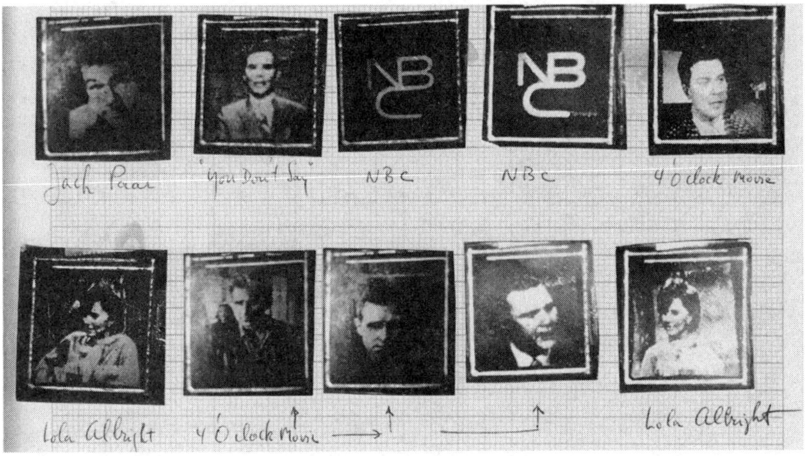

Figure 3.7. Screenshots from John van Raalte's electron-beam-addressed LCD. These were some of the earliest television broadcasts viewed on a liquid crystal display. (Van Raalte, Notebook 32465, 77. David Sarnoff Library Collection, courtesy of Hagley Museum and Library.)

temperature."[151] Those discussions had been largely speculative, but following the successful synthesis of room-temperature liquid crystal mixtures, Lechner and half a dozen other CRL engineers would have a chance to put their ideas into practice.

They began by working with Zanoni to construct an "exerciser," a small display with two rows of eighteen elements that could simulate addressing at standard television rates (fig. 3.8).[152] The major challenge bringing this display online was the same one that Van Raalte and Moles had identified a year earlier: isolating each liquid crystal picture element so that signals directed at a particular x-y intersection would not leak over to cells in neighboring rows or columns. They had suggested that a single diode placed in series with each picture element would eliminate this problem, but two of Lechner's engineers, Frank Marlowe and Juri Tults, confirmed this "classical method" would result in turn-on delays and low brightness.[153]

In response to this information, Lechner and his CRL colleagues designed two improved addressing circuits, referred to as double-diode capacitor (D^2C) and field-effect transistor capacitor (FETC),

Figure 3.8. Bernard Lechner utilized this liquid crystal "exerciser" to develop the first active matrix addressing system. The circuitry he designed would later be adapted for use in today's televisions, laptop computers, and smartphones. (David Sarnoff Library Collection, courtesy of Hagley Museum and Library.)

based on the equipment needed to isolate each picture element.[154] After testing these designs on the LCD exerciser using discrete circuit components, Lechner asked colleagues in the DSRC's new Integrated Circuit Center and RCA's semiconductor division in Somerville, New Jersey, for assistance fabricating two 1,200-element displays, one for each addressing system. These prototypes would be tiny, measuring slightly over three-quarters of an inch diagonally, but each would be fully integrated, capitalizing on the latest advances in solid-state manufacturing, including Paul Weimer's thin-film transistors.[155]

By taking this step, Lechner hoped to streamline the fabrication of future displays, and in fact, today almost all LCDs—from big-screen televisions to super-slim cell phones—utilize what is now known as an *active matrix* addressing system derived from his FETC circuit.[156] (The term refers to the presence of an "active" switching element, like a transistor, behind each pixel, rather than a "passive" set of electrodes.) At the time, however, as technical difficulties ranging from poor diode yields to broken silicon wafers piled up, his updates took on a more pessimistic tone. "Both of these approaches [D^{2}C and FETC] represent ambitious undertakings in semiconductor technology and the progress has been slow," Lechner summarized in the spring of 1968.[157] The group's increasing reliance on solid-state researchers who did not answer directly to anyone in the liquid crystal group further aggravated matters.[158] As Heilmeier and Zanoni constructed simpler devices, such as electronic windows and numeric counters, Lechner postponed the expected completion of his group's 1,200-element displays until one April day in 1968, when he received a memo from RCA's public relations department. Apparently, the corporation was ready to share the LCD with the world.[159]

DUELING CONCEPTIONS OF INDUSTRIAL RESEARCH

The walls of silence surrounding the RCA liquid crystal project had started to crumble in November 1967, when Heilmeier, Barton, and Zanoni jointly authored a technical report on dynamic scattering. Unlike their red-covered progress updates, this document was accessible to the entirety of RCA's technical staff and was later reworked into an article submitted to the *Proceedings of the IEEE* in January 1968.[160] The DSRC's annual report for 1967, published soon thereafter, also included a section on LCDs and the first company-wide references to dynamic scattering.[161]

What prompted these revelations after more than two and a half years of secrecy is not immediately evident from the group's internal reports. Compared with the chemists' synthesis of room-temperature liquid crystals in the spring of 1966 or Heilmeier's formulation of the momentum exchange model for dynamic scattering that November, the research conducted in the final quarter of 1967 was transitional in nature. The development of new nematic mixtures, improved theoretical frameworks, and superior addressing circuitry all furthered the group's strategic goals, but none of them, in and of themselves, suggested the time had come to take the LCD public.[162]

Ironically, the event that triggered the dissemination of information about the LCD may not have come from within Heilmeier's group at all. Some credit must also be given to Richard Williams, whose liquid crystal patent—submitted in November 1962—was finally issued in May 1967. The prototypes that Williams and Larach demonstrated bore little resemblance to Heilmeier and Lechner's sophisticated displays, but the former provided a foundation on which RCA could build its LCD patent portfolio. By the middle of 1968, Heilmeier's team had filed applications for more than a dozen additional patents covering all aspects of LCD manufacturing, allowing them to share their accomplishments without undermining the future commercialization of the technology. As one RCA patent counsel noted, "we expect to obtain broad coverage on DSM [dynamic scattering mode] devices" by the early 1970s.[163]

Williams set the stage for RCA's future LCD initiatives, but by the time he received his patent, he had withdrawn to the margins of the DSRC liquid crystal project. Heilmeier occasionally consulted him about the molecular behavior underlying dynamic scattering, but his research agenda remained as varied as ever, including investigations into the surface chemistry of semiconductors and solid-state electron multiplication.[164] For him, this freedom to multitask was the most appealing aspect of working in a corporate environment. Compared with his counterparts in the academy, who constantly squabbled with faculty members and administrators for space or resources, Williams enjoyed a "good bit of autonomy."[165] "Not autonomy to spend a lot of money," he emphasized. "I couldn't tell five technicians to go do something, but I could do a lot myself, and so I was accustomed to doing things by myself. And so I was given a good bit of freedom, and that was the way I liked to work."[166]

Unsurprisingly, Heilmeier shared Williams's preference for the corporate laboratory over the ivory tower. "University people aren't as free as they think they are," he explained in a 1969 interview:

> Suppose two people start graduate school with the same qualifications. One does all his work in the university, while the other does his course work in class but his laboratory work in an industrial lab. Five years later, the

industrial fellow usually will be much more productive in terms of patents, publications and conference presentations.[167]

Yet while both Williams and Heilmeier explained their career choice in terms of personal freedom, the handling of their respective liquid crystal display projects reveals that each possessed a very different understanding of what that autonomy signified within the context of the DSRC. For Williams, autonomy meant retaining the ability to engage in numerous scientific investigations; for Heilmeier, it represented an opportunity to follow a single research pathway wherever it might lead.

This difference in perspective calls to mind Isaiah Berlin's 1953 essay "The Hedgehog and the Fox." Quoting the Greek poet Archilochus, Berlin distinguished between two types of thinkers: the fox, who "knows many things," and the hedgehog, who "knows one big thing."[168] Berlin applied these categories to literary figures, but historian Thomas Hughes later extended them to technological system builders like Thomas Edison, whom he classified as hedgehogs because of their capacity to link all aspects of their business (invention, management, finance, etc.) to a central vision.[169] The example of liquid crystals suggests the relevance of this methodology to industrial scientists and engineers. Some members of RCA's technical staff, like Williams, were foxes, pursuing several investigations at a time without being disheartened if any one of them did not pan out. Others, like Heilmeier, were hedgehogs, viewing all of their investigations as pieces of a unified research program and committing all of their energies to advancing that agenda.

Embracing Berlin's taxonomy adds a personal dimension to the discovery narratives presented at RCA's 1968 press conference. On that occasion, Hillier's description of the LCD's origins conformed to the general parameters of the linear model. Basic research into the fundamental properties of materials revealed a new electro-optic effect that made possible a wide range of practical applications. In the case of RCA's liquid crystals, this process actually occurred twice.

Following two very different paths, Williams and Heilmeier each reached similar conclusions about the display potential of nematic materials. The subsequent divergence in their investigative trajectories confirms the extent to which an individual's research style shapes the outcome of scientific projects, even in the supposedly impersonal confines of a corporate laboratory.[170]

An industrial research facility like the DSRC benefitted strongly from harboring both foxes and hedgehogs under its roof. Williams's imaginative decision to expand his study of electric fields and absorption spectra to encompass solid and liquid crystals resulted in the demonstration of new electro-optic effects with possible display applications. By the same token, his vulpine tendencies and preference for solitary research left him poorly disposed to mount an extended campaign on behalf of the LCD. Such an endeavor would be a major diversion from his other research activities. On the other hand, Heilmeier, the quintessential hedgehog, approached the same physical phenomena—domain formation and the turbulence that would become known as dynamic scattering—as the latest stages of his ongoing search for an improved laser modulator. This notion drove him forward, leading to the observation of new effects, such as electronic color switching, and the belief that liquid crystals could serve not merely as modulators but full-fledged displays.

More than any other factor, Heilmeier's dedication to that objective underlay the organization and success of RCA's liquid crystal program. Others, including Williams and Larach, embraced the general concept of the LCD but never moved beyond workbench demonstrations. Kleitman championed the project early on, capitalizing on his managerial status to pull strings that would have been inaccessible to Williams or Heilmeier, but his involvement diminished over time.[171] Heilmeier stepped into the leadership vacuum, expanding the group's membership to include more chemists and physicists while forging close bonds with SRL and CRL engineers possessing previous flat-panel experience.

Heilmeier and the research team he oversaw between 1965 and 1968 were vital to liquid crystals' emergence as a strategic priority within the DSRC. Press releases and academic articles presented only partial details concerning this group's membership and achievements, but all of them reaffirmed the central role of RCA scientists and engineers in advancing this new technology from inception to prototype. It remained to be seen whether or not that experience would facilitate an equally smooth transfer of their displays from Princeton to RCA's operating divisions. The organizational and technological obstacles confronting the DSRC liquid crystal group did not disappear once its work became public knowledge. If anything, they multiplied, returning in new and unpredictable forms as Heilmeier's team fought to move dynamic scattering into the marketplace.

4 DISRUPTIVE DISPLAYS, 1968–1971

An explosion of colors reminiscent of fireworks graced the cover of RCA's annual report for 1968. This document marked the first instance in four decades that David Sarnoff refrained from writing an introductory message recounting the past year's achievements to his company's shareholders. A severe case of shingles had struck the seventy-seven-year-old executive the previous summer, and the weakened businessman had ceded this responsibility to his son and designated successor, Robert Sarnoff.[1]

Despite an initial reluctance to enter the family business, the younger Sarnoff had taken a job in RCA's sales department in 1948. By 1953, he had been promoted to executive vice president of NBC, and two years later he became head of the network. The swiftness of his ascent lent credence to internal rumors that the General was grooming Robert to take command of RCA, and few were surprised by the 1965 announcement that the latter would replace Elmer Engstrom as the corporation's president.[2] Following his father's illness, he would also be named RCA's chief executive officer.

Robert Sarnoff had been in power for only a short time, but the 1968 report confirmed that he had already made his mark on his father's electronics empire. In a literal sense, he had replaced the company's familiar logo—a circle containing the letters "RCA"

Figure 4.1. Robert Sarnoff unveils the new RCA logo in the corporation's 1967 annual report. (RCA Corporation, *Annual Report 1967*, 5. David Sarnoff Library Collection, courtesy of Hagley Museum and Library.)

underlined with a lightning bolt—with the blocky monogram that has persisted under various owners to the present day (fig. 4.1).[3] This move, along with the decision to refer exclusively to "RCA" in official documents—not the "Radio Corporation of America"—signaled his belief that the firm should expand its activities beyond telecommunications.

Embracing the strategy of expansion through acquisition that his father had pioneered with the 1929 purchase of the Victor Talking Machine Company, Robert Sarnoff would reinvent RCA as a conglomerate.[4] Under his watch, RCA broadened its traditional focus on consumer electronics by purchasing firms such as Random House Publishing and the Hertz Corporation. "The word that best characterizes the modern RCA is diversity," Sarnoff explained in the 1968 shareholders' report.[5] To illustrate this point, he referred readers to an accompanying twelve-page foldout portraying "a cross section of the 12,000 products and services created or marketed by RCA at home or abroad."[6] One had to marvel at the sheer variety of items represented. A single page showcased the cover of a John Updike novel, a Henry Mancini album, and NBC's Captain Kirk and Mr. Spock posed within phaser range of a superconductive magnet, a VHF transceiver, and an RCA mainframe computer.

Lest his shareholders worry that RCA had grown overly reliant on its existing businesses, Sarnoff also included a section in his letter describing "a significant scientific effort in 1968" aimed at creating "the foundation for some promising new enterprises in the decade ahead."[7] Foremost among these was "a revolutionary type of electronic display, based on research in liquid crystals."[8] The fantastic colors on the report's cover demonstrated the capacity of these substances to modulate the passage of light. Borrowing his father's mantle of technological prophecy, Sarnoff proclaimed that liquid crystals "could lead to an entire new range of electronic display products—from wrist watches to pocket-size television receivers that could be viewed in broad sunlight."[9] Later on, DSRC managers expanded on these claims, suggesting LCDs might be used in "instrument displays for automobile dashboards and airplane cockpits, scoreboards, stock tickers, and, perhaps someday, flat screen television receivers."[10]

Faced with these glowing predictions, RCA shareholders could be forgiven for expecting additional updates about this new technology, but the next year liquid crystal displays vanished from RCA's annual reports. At the precise moment when the LCD left the laboratory, company bulletins withheld mention of any liquid crystal investigations underway at the DSRC. They also avoided any reference to the formation of a new liquid crystal R & D group at RCA's semiconductor division or the inauguration of the world's first LCD manufacturing operation.

The silence surrounding RCA's liquid crystal displays between

1968 and 1971 stands in sharp contrast to the fanfare of their public premiere. Granted that the more mundane aspects of product development did not lend themselves well to promotional efforts, it is nevertheless striking how abruptly RCA's management changed its stance on LCDs and the researchers responsible for their creation. The scientists and engineers previously hailed as exemplars of interdisciplinary collaboration became personae non gratae.

They were not alone. Within RCA and across the United States, projects rooted in speculative research were coming under increased scrutiny in the late 1960s. After pouring money into fundamental science for decades, industry leaders and policy makers started to question the return on their investments. Managers at GE, DuPont, and Kodak worried that their massive laboratories were not generating as many blockbuster products as expected.[11] Further fueling their apprehension was Project Hindsight, a Defense Department study that concluded in 1966 that basic research played only a minor part in the development of twenty major weapons systems. Even as the National Science Foundation raced to rebut Hindsight's findings, confidence in the linear model warped into fractals of doubt rippling across all levels of the nation's research establishment.[12]

Every company responded to this crisis of faith differently. At RCA debates over the contributions of the Princeton laboratories were nothing new, but when push came to shove, researchers could always count on David Sarnoff's protection. Under his son, the DSRC's privileged position was no longer guaranteed, and what we might refer to today as "disruptive" technologies—including liquid crystal displays—faced increased opposition from RCA's headquarters and operating divisions. George Heilmeier's team responded to these circumstances as their flat-panel predecessors had, seeking out new partners and resources to support their cause, but in this instance those tactics failed to prevent the marginalization of the LCD.

Years later, Heilmeier attributed the declining status of liquid crystal research to short-sighted leadership at RCA's headquarters and operating divisions that squandered their first-mover advantage.[13] There is some truth to that assessment, but placing blame solely on management risks casting members of the firm's technical staff as passive recipients of corporate R & D policy. In actuality, company decision makers continued to rely heavily on RCA scientists and engineers for guidance as they considered possible LCD applications. The results of these discussions may not have lived up to researchers' early expectations but reflected their recognition of dynamic scattering's limitations as well as their fading confidence in RCA's ability to commercialize liquid crystal displays.

FINDING A NICHE AT THE DSRC

As RCA's shareholders opened their mailboxes to discover the company's glossy, full-color 1968 report, managers at the firm's operating divisions received a package of their own: a 186-page tome summarizing the research projects underway at the DSRC. This dense volume also cited liquid crystals prominently in its introduction, focusing less on possible LCD applications and more on Heilmeier's research team. "The liquid-crystal work started several years before in some quite basic research," the DSRC report noted. "But its successful public demonstration was brought about by interdisciplinary applied research involving physicists, chemists, and electronic engineers working and communicating together."[14]

For the DSRC's managers, the LCD was evidence of both the linear model's validity and the Princeton laboratory's value as a space for creative collaboration. Such arguments were unnecessary when David Sarnoff ran RCA. Even after stepping down as CEO, the General remained a frequent visitor to his namesake research center. He kept an office in Princeton and spent much of his retirement supervising the construction of a new wing in which to house his personal papers.[15]

Robert Sarnoff possessed neither his father's technical background nor his connections to Princeton. A Harvard graduate, Robert Sarnoff had obtained the formal education that poverty denied his father but found the public relations side of the electronics business more appealing than the inner workings of radios or televisions. As president of NBC he oversaw the network's transition to color broadcasting and showed a keen eye for marketing strategy.[16] For DSRC personnel, the open question in 1968 was whether these qualities would translate into the effective oversight of a corporate research operation.

Early on it seemed as though the younger Sarnoff would maintain the status quo in Princeton. Out of deference to his father, he offered the research center an exemption from replacing "The Radio Corporation of America" on its signs with the new RCA trademark. This gesture may have appeased the General, who expressed fondness for the company's original name and "meatball" logo, but the technical staff remained wary of their new CEO's policies. Many wondered what his calls for diversification meant for the firm's R & D budget. The 1968 DSRC report promised that RCA's new emphasis on "satisfying consumer wants profitably rather than just selling products" would not result in "any diminishing of the quality or quantity of RCA Laboratories' more basic and high-risk exploratory research."[17] Yet it also alluded to new screening procedures to select "viable and profitable" projects, suggesting an increased need to justify speculative investigations.[18]

One figure who discerned the vulnerability of the DSRC's fundamental research programs was the laboratory's leader, James Hillier. Hillier had previously moderated John Burns's attempts to reorient all work underway in Princeton toward application development and encouraged the revival of long-term projects once color television revenues replaced patent royalties as a source of research funding. But this success, he realized, was only temporary. Once the color TV windfall ran its course, RCA's technical staff would again find itself at the mercy of outsiders unless they established that their work was vital to the company's future. As Hillier put it, "the rest of the corporation would have to see the laboratory as a life raft not an albatross."[19]

Hillier sought to shore up the DSRC's position through the 1964 formation of the Interim Research Planning Committee (IRPCO). IRPCO's mission was to classify all ongoing research at Princeton according to its relevance to RCA product lines with the aim of reducing projects with no obvious commercial significance. Functionally, it resembled the panel of experts Burns had recruited to evaluate the DSRC shortly after taking office, but Hillier believed this in-house oversight would be more palatable than incursions from the operating divisions.[20] IRPCO did win over many veteran engineers who had grown frustrated with the DSRC's drift toward the theoretical over the previous decade. Others, who were now accustomed to a high degree of intellectual freedom, resisted the new guidelines. The result, in the words of one observer, was "a lot of friction and a damn near technical revolt."[21]

George Heilmeier opposed IRPCO's agenda, which threatened to cut off support for his liquid crystal research team before it achieved any tangible results. Therefore, he probably welcomed the news that the committee's reforms would lapse following Hillier's sudden promotion to corporate vice president at the end of 1968.[22] If so, he soon learned that rumors of IRPCO's demise had been somewhat exaggerated. Hillier's replacement, William Webster, had "tried to get things directed around towards more immediate needs in the divisions" since becoming head of the DSRC's Electronics Research Laboratory in 1959.[23] Now, as he wrote in the research center's 1969 annual report, the distinguishing trend of work carried out under his watch would be "increased relevance to RCA's needs" even if that required "some sacrifice of fundamental research in favor of advanced development."[24]

By that point, Heilmeier's team had won praise from DSRC managers, including Hillier and Webster, for their dynamic scattering prototypes.[25] All the same, it was unclear how RCA's looming strategic realignment would affect the LCD project, which had always straddled the line between fundamental and applied research. To avoid being branded as overly speculative, Heilmeier and his colleagues resorted to three strategies meant to link even more theoretical liquid crystal investigations to the generation of new products.

The first approach, pursued primarily by technicians such as Louis Zanoni and Ronald Friel, consisted of designing entirely new applications for dynamic scattering LCDs. Zanoni had already mastered

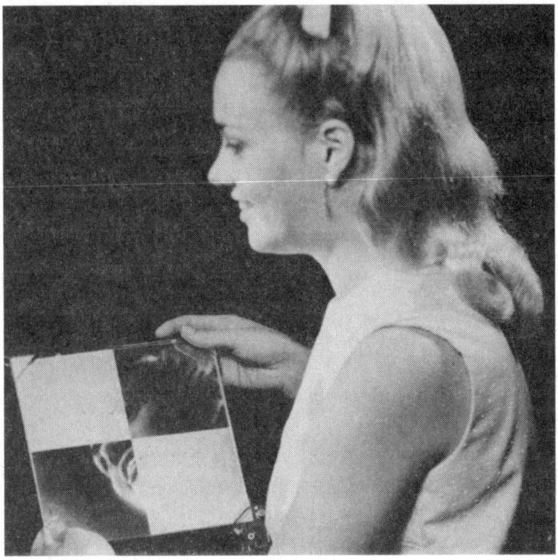

Figure 4.2. Lynn Thomas of the DSRC's Consumer Electronics Research Laboratory examines her reflection in a dynamic scattering mirror built by Louis Zanoni for a 1969 art exhibition at New York City's Wise Gallery. ("Mirror, mirror . . . ," *RCA Radiations* 16, no. 4 [July–Aug. 1969]: 2. David Sarnoff Library Collection, courtesy of Hagley Museum and Library.)

the generation of high-resolution still images using liquid crystals. Partnering with Friel and semiconductor specialists Robert Lohman and Steven Hofstein, he now expanded his repertoire to construct a battery-powered voltmeter and both digital and analog LCD clocks.[26] (As seen earlier, Lohman posed for a photograph alongside the digital clock, which he demonstrated at RCA's 1968 press conference.) Typically, the two technicians delivered their displays directly to the operating divisions, though they sometimes presented them at RCA marketing meetings.[27] In one case, they took their handiwork public, collaborating with nuclear engineer turned sculptor Earl Reiback on a June 1969 exhibition of technological art in New York City (fig. 4.2).[28]

Along with these new prototypes, liquid crystal researchers also contemplated the incorporation of LCDs into existing RCA products. Initially the group held out hope of fulfilling its primary objective "to replace CRT's for commercial TV and other display applications," pointing to Bernard Lechner's exercisers as proof of a dynamic scattering television's technical feasibility.[29] Lechner pushed forward with that project until the end of 1968, when, citing "painfully slow" progress and "formidable problems with integrated circuitry," his group ceased work on their 1,200-element displays.[30] Instead, they considered adapting liquid crystals for use in the DSRC's Electrofax photocopier and Homefax facsimile projects. Each of these relied on photosensitive paper, the former to reproduce documents and the latter to generate printed material coordinated with specific television broadcasts (e.g., a recipe to accompany a cooking show). Lechner's liquid crystal arrays, though too small for television, might be effective light shutters in these systems.[31]

Before setting aside his matrix-addressed LCDs, Lechner drew on the experience of his previous electroluminescent display project and availed himself of the final strategy adopted by DSRC liquid crystal researchers: applying for a government contract.[32] Before 1968, the shroud of secrecy over dynamic scattering eliminated the possibility of securing external funding. Now that the LCD was public knowledge, RCA's technical staff could approach agencies less closely wedded to short-term payoffs than the DSRC to finance their liquid crystal studies.

Lechner's attempt to sustain the development of an LCD television was unsuccessful, but RCA liquid crystal researchers received over $500,000 in research funding from military and civilian agencies between 1968 and 1972.[33] The US Air Force and NASA were especially eager to develop thin, lightweight displays for use in flight instrumentation and sponsored fundamental research contracts on the synthesis of new mesomorphic compounds and their behavior under applied fields.[34] External funding also provided opportunities to reconsider electro-optic phenomena that had not garnered as much notoriety as dynamic scattering. The Air Force Materials Laboratory in Dayton, for example, expressed interest in mixtures of nematic and cholesteric materials that exhibited a storage effect, remaining opalescent even after a triggering field was removed.[35] Similarly, NASA's

Electronics Research Center renewed work on the guest-host effect in hopes of creating flat-panel color screens for future space capsules.[36]

Through this tripartite strategy, Heilmeier and his colleagues were able to sustain their liquid crystal research between 1968 and 1970. In taking this step, they followed the example of the DSRC's leadership. Much as Hillier and Webster recast Princeton's contributions as vital to an increasingly diversified RCA, the liquid crystal team wished to secure a niche for themselves within the ever more profit-minded DSRC. Both groups wanted to strike a balance between the freewheeling experimentation of the "building block" era and the focused product orientation of the post-IRPCO period and to do so on their own terms. Whether or not the DSRC as a whole would succeed at this task remained uncertain in the final years of the 1960s, but in the near term, RCA's liquid crystal group felt confident enough to contact the operating divisions and discuss the future of dynamic scattering.

SEARCHING FOR ALLIES IN SOMERVILLE

In September 1967, Lawrence Murray, the head of the device physics group at RCA's Electronic Components Division (ECD) in Somerville, New Jersey, invited associate engineer Richard Klein to accompany him to a meeting in Princeton.[37] Klein had joined ECD, which oversaw the company's semiconductor development efforts, a year earlier and became involved in an army-sponsored project to build digital readouts using light-emitting diodes (LEDs).[38] Klein's group kept in regular contact with LED researchers throughout the corporation, so a drive to the DSRC was not unusual. Still, on this occasion Murray guaranteed him that they were going to see something new.[39]

Sure enough, upon arrival at the DSRC, Klein and Murray were escorted into a room and introduced to George Heilmeier and Louis Zanoni. Heilmeier led the visitors toward a workbench and presented them with a pane of glass connected to a power supply. Careful inspection revealed hints of a picture etched into the material, but otherwise there was nothing out of the ordinary about this experimental setup. Then Heilmeier flipped a switch and a familiar black-and-white image appeared on the previously transparent square. "It was a TV test pattern," Klein remembered. "And you know the thing pops up, and I almost fell over!"[40]

Once Klein had recovered from this surprise, Heilmeier provided a brief introduction to dynamic scattering and disclosed that the Princeton group wanted Somerville's assistance turning their LCD prototypes into products. The DSRC had already reached out to Somerville to fabricate integrated addressing circuitry for Lechner's matrix displays, but Heilmeier now wished to lay the groundwork for a more sustained partnership between the two organizations, culminating in the establishment of an LCD pilot plant. Murray's group, which had experience dealing with the electro-optic behavior of new materials, seemed an ideal avenue to introduce liquid crystals to a division more comfortable with silicon. Klein enthusiastically embraced the idea. "To me, liquid crystal was a gimme. I mean, it was a gimme!" he explained. "It had to go because it was the only technology that really matched up well with integrated circuits. It was very low power. It could be scaled up in size. It didn't fight the light in the environment; it used the light in the environment!"[41]

Klein's LCD experiments began in relative isolation with nothing but a handful of liquid crystal samples from the DSRC and assurances from his ECD superiors that additional personnel and funding were forthcoming. The former promise, at least, turned out to be accurate. In early 1968, Sandor Caplan, another member of Murray's LED project, started working on liquid crystals too. He became responsible for evaluating the performance characteristics of dynamic scattering displays, measuring "things like operating voltage, operating frequency, temperature range of operation. We didn't do any shock or vibration [testing] particularly, but stuff like that."[42] Though reliant on the DSRC for chemicals and equipment, Klein and Caplan's team eventually grew to include chemist Howard Sorkin, physicist Arthur Elsea Jr., engineers Herman Stern and Henry Schindler, and manufacturing expert George Hiatt.[43] Together this group worked to standardize procedures for the large-scale assembly of liquid crystal displays.

As much as the ECD personnel respected their Princeton colleagues' ingenuity, the prototype devices Heilmeier had shown the public in 1968 were nowhere near ready for the marketplace. Robert Lohman later remarked that the team's digital clock "lasted just long enough to take a picture."[44] With the exception of the basic "sandwich cell" configuration underlying all dynamic scattering displays, every aspect of LCD construction underwent a major transformation in the hands of the staff at Somerville. The DSRC team, for example, had taken to constructing displays using synthetic quartz substrates, which retained a smoother surface than normal glass after the application of chemical coatings but which were also substantially more expensive.[45]

The techniques they had developed to fill and seal their displays were similarly impractical. Klein summarized how both of these processes were performed before ECD's involvement:

> First of all, they'd saw two little grooves parallel along the sides, okay? Then you'd take a piece of one mil Mylar, lay it on four sides. Put the sandwich together. Then, you'd take a hypodermic needle, slide it along the groove and inject the liquid crystal, which would osmose across the space, and the excess would wind up in the two little grooves. It wasn't sealed. It was clamped.[46]

Over time, the Somerville group resolved each of these issues. Caplan procured a material developed for use in fighter jet canopies by Pittsburgh Plate Glass that was similar in texture and electrical behavior to synthetic quartz but could be purchased at lower cost.[47] Meanwhile, Klein, Schindler, and Stern experimented with a variety of epoxies to hold the glass plates together. The resulting devices were sturdier than Princeton's clamped setups, but the ECD researchers soon learned that chemical reactions between the liquid crystals and certain adhesives "poisoned" the displays, rendering them nonoperational.[48] They eventually overcame this issue with frit—a mixture of ground glass and binders that RCA had used to form hermetic glass-to-glass and glass-to-metal seals in CRTs.[49] The group also replaced the injection filling method, which had complicated the uniform distribution of dopants that optimized display performance and relied on grooves with an annoying tendency to clog. In its place, they drilled two small holes in opposite corners of the cell and used a vacuum to draw liquid crystal mixtures across in a uniform layer.[50]

As Caplan and Klein refined their display construction techniques, other personnel considered the problem of scaling up to assembly line production. Earlier there had been no pressing need for large batches of the room-temperature liquid crystal mixtures that Joseph Castellano and the other DSRC chemists had developed. Now Howard Sorkin supervised the expansion of Somerville's chemical production facilities. At the same time, George Hiatt consulted with staff in Princeton and Somerville to divide the LCD fabrication process into the discrete steps required on a factory floor. By 1969, the new LCD operation had grown so extensive that a move was in order. "There was no space available for, you know, such a gigundic [*sic*] undertaking at Somerville," Caplan recalled.[51]

Consequently, the entire project was relocated to a company-owned warehouse in Raritan, a few miles away from the main ECD building. The Raritan site served as an all-purpose advanced development space, housing everything from an optoelectronics group working on laser diodes to a small publishing operation.[52] Now it would become home to RCA's new LCD manufacturing facility as personnel in Princeton, Somerville, and New York started debating what its first products should be.

IN PURSUIT OF A PRODUCT

Auspicious as the move to Raritan might have been, the fledgling liquid crystal operation still lacked a coherent product-development strategy. ECD had made good on its promise to increase the number of people assigned to the LCD project but appeared reluctant to expand its budget until the group could demonstrate sufficient consumer demand for the new displays. Fortunately, the ECD group did not have to accomplish this task on their own. Before leaving Somerville, their efforts had attracted the attention of a marketing executive from RCA's tube manufacturing division in Harrison, New Jersey, named Jack Riddel, who, Caplan recalled, "took a shine to this liquid crystal thing unbelievably."[53] Riddel chatted with ECD personnel and Heilmeier's team at the DSRC about possible LCD applications before mobilizing his network of sales contacts.[54]

By the time ECD set up shop in Raritan, Riddel had secured three product-development contracts. ("Jack was a real operator," Heilmeier told interviewers later.[55]) The first came from a New York–based firm called Ashley-Butler, which offered RCA $100,000 to construct an animated point-of-purchase advertising display. Veeder-Root, a leading producer of gauges and mechanical tabulation devices, matched that sum in exchange for a liquid crystal readout for gasoline pumps. Finally, Jervis Corporation supplied $50,000 for an automobile rear-view mirror that used dynamic scattering to eliminate nighttime headlight glare.[56]

With a quarter million dollars at stake, the ECD engineers sprang into action. The combined value of Riddel's contracts paled in comparison to the corporation's $271 million R & D expenditures for 1969, but both Princeton and Raritan viewed their fulfillment as a stepping stone toward broader investment in LCD technology.[57] All three projects drew on preliminary research conducted at the DSRC, but each needed additional refinements to meet their customers' specifications. In some instances, these modifications were relatively straightforward, such as adding colored films to the gas pump displays to provide higher contrast.[58] Others required more ingenuity. Caplan's design for a day-night mirror worked well indoors, for example, but the slow response of the liquid crystals at low temperatures led to a collaboration with Klein to devise a self-regulating heater.[59] Similarly, although Heilmeier and Zanoni had discussed the construction of simple animated LCDs, determining the electrode layout for such a device and a mechanism to automatically activate images in the proper sequence was left to the staff at Raritan (fig. 4.3).[60]

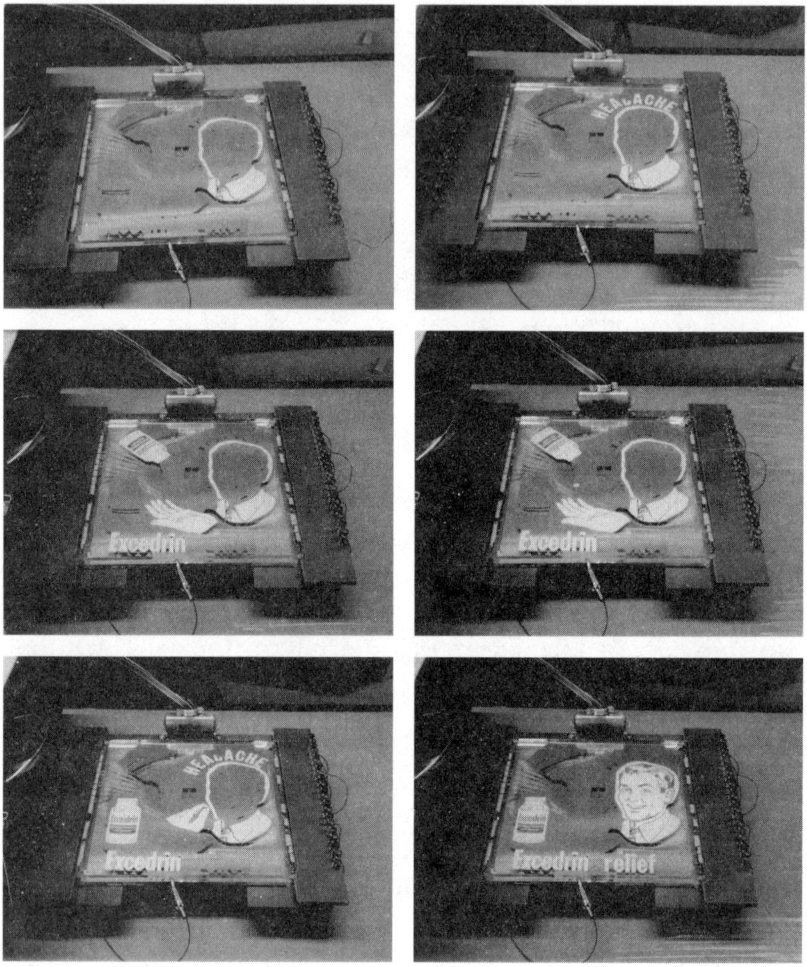

Figure 4.3. Richard Klein, an engineer at RCA's semiconductor division, devised a mechanism for sequentially activating different portions of a dynamic scattering LCD and incorporated that system into this prototype point-of-purchase advertising display. (David Sarnoff Library Collection, courtesy of Richard Klein and Hagley Museum and Library.)

These technical feats may have impressed their clients, but RCA's continued reliance on external funding raised concerns about its commitment to liquid crystals. As the ECD team moved closer to bringing their pilot plant online, its members wondered whether the company had any plans for the technology besides the trio of contacts Riddel had lined up for them. To Klein it seemed "nobody in senior management who had any insight ever looked at this or ever talked to people and said, as I think you should, 'What kind of products can you make?'"[61]

Actually, in September 1969, RCA's marketing staff initiated a six-month study to answer precisely that question. Working alongside a California market research firm, the company organized a pair of brainstorming sessions at the New York Hilton in order "to identify the maximum number of applications for the liquid crystal dynamic scattering phenomenon."[62] These conversations included representatives from RCA's headquarters and operating divisions, with George Hiatt and Jack Riddel attending on behalf of ECD. Participants were encouraged to propose as many LCD applications as they could without worrying about their technical or economic feasibility. The resulting list contained more than 150 potential uses for dynamic scattering, including everything from automobile speedometers to zoo enclosures for nocturnal animals.[63] The marketing team solicited additional feedback from RCA's advanced-development groups in Camden and Indianapolis and several sources outside the corporation.[64] They also asked Heilmeier to submit a written forecast of "how the liquid crystal technology may be expected to evolve."[65]

After sifting through this information, the marketers presented their conclusions in March 1970.[66] In their report they acknowledged that dynamic scattering had "commercial significance as the enabling technology for several types of products," but they argued that the company's idea-generating strategies had "failed to identify any <u>immediate</u> opportunities for <u>new end products</u> based on liquid crystals."[67] There were several products on the market, such as desktop calculators, that could be enhanced through the use of LCDs, and within a decade it might be possible to produce "a liquid crystal cathode ray tube substitute" matching the performance of existing black-and-white sets.[68] ("The advent of a color liquid crystal CRT device," the report noted, "cannot be forecast."[69]) But among those products available in the near future, almost all had "relatively small market potential, except alphanumeric displays, point-of-purchase advertising signs, and flat glass glazing products."[70] Not coincidentally, these three categories corresponded precisely with the projects under development at Raritan.

In retrospect, one item was conspicuously absent from this list of recommendations: the electronic wristwatch. In the 1965 paper outlining his eponymous law, chemist Gordon Moore—the director of R & D at Fairchild Semiconductor—had observed that only the lack of a display that could be driven by integrated circuits prevented the construction of such a product.[71] Perhaps the low power requirements of liquid crystal displays, which reflected light rather than emitting it, offered a solution to this puzzle. Hillier directly referenced the concept at the LCD group's 1968 press conference, commenting that "it doesn't take too much imagination to foresee an all-electronic wrist watch being a few years down the pike."[72] After seeing the dynamic scattering clock in action during that event, domestic and international watchmakers contacted RCA about the prospect of designing a wearable LCD timepiece. The company's marketing study dismissed the idea. Digital and analog clocks could be built with existing dynamic scattering displays, but an "electronic watch based on integrated circuits, with a liquid crystal readout" would have to wait because of "an apparent incompatibility in the operating voltages of the watch circuit and the display."[73]

Although this conclusion resulted from conversations between RCA's marketing staff and members of the company's liquid crystal operation, the two groups viewed the voltage incompatibility problem very differently. Where the marketers saw it as a bar to further action, personnel in Princeton and Raritan recognized yet another in a long line of issues that could be solved if RCA committed sufficient resources. DSRC engineer Steven Hofstein had already filed a patent application for an LCD wristwatch in connection with his work on the

Figure 4.4. Not functional timepieces, but a remarkable simulation! These LCD watches, assembled by personnel at RCA's Raritan facility, were powered by hidden wires and contained no batteries or integrated circuits. (David Sarnoff Library Collection, courtesy of Richard Klein and Hagley Museum and Library.)

dynamic scattering clock.[74] Others in Princeton believed that new research into more efficient CMOS (complementary metal-oxide-semiconductor) integrated circuits would make such an invention practical.[75] Their ECD colleagues supported the plan, going so far as to construct a series of mock-up digital watch displays (fig. 4.4).[76] Together their efforts may well have persuaded the marketers and managers in New York City to reconsider their analysis had they not faced obstruction from an unexpected source.

When the ECD liquid crystal project began, Klein and Caplan reported to Lawrence Murray, who granted them a significant amount of freedom in their day-to-day work. This state of affairs changed abruptly at the end of 1968. As the group prepared to move to Raritan, the corporation decided that the new facility required a more formal organizational structure. In place of Murray and his laissez-faire management style, leadership of the ECD operation passed to Norman Freedman, a chemical engineer who had distinguished himself in the 1950s by devising a process to apply precisely thousands of phosphor dots to the faceplates of color television picture tubes.[77] More recently, he had supervised the organization of superconductor laboratories at the DSRC and RCA's operating division in Harrison.[78] Now it was hoped that Freedman's manufacturing experience would ensure the successful establishment of an LCD assembly line.

As his subordinates soon learned, Freedman was a forceful manager who insisted on retaining complete control over all aspects of his projects. This philosophy may have been effective with a product such as color television, whose significance to the corporation was above reproach, but it risked alienating personnel whose support was needed to nurture a largely unknown technology like the liquid crystal display. Freedman refused, for example, to allow collaboration between the LCD facility in Raritan and the main semiconductor division in Somerville even though it was obvious that dynamic scattering applications would benefit from closer coordination with integrated circuit designers. "We were told explicitly, 'You may not talk to anybody in integrated circuits,'" Klein explained. "And we talked to our friends in integrated circuits under the table, and they were told by their managers, 'You are not to cooperate with these guys.' And it was because they all hated Norm."[79] Freedman's ban on collaboration was only one means by which he consolidated power. Along with ceramic engineer Walter "Lucky" Lawrence (another veteran of RCA's Harrison operation), he also imposed order on the ECD group through new regulations, such as the replacement of monthly progress reports with weekly updates and the promotion of personnel recruited from outside the operation over veterans such as Klein and Caplan.[80]

Freedman's resistance to interdivisional collaboration and sidelining of some of his more experienced engineers effectively paralyzed the ECD liquid crystal program. Raritan's leadership confined its manufacturing activities to numeric counters, point-of-purchase displays, and dynamic scattering mirrors, none of which required substantial collaboration with either Princeton or Somerville. The DSRC team, which dedicated "roughly half" of its research effort in 1969 "to direct support of the liquid-crystal products activity at EC [Electronic Components] in Raritan," found its ideas rebuffed with increasing frequency.[81] From their perspective the ECD operation was being controlled by individuals with no understanding of dynamic scattering's potential. Heilmeier later accused ECD personnel of rejecting the digital wristwatch "because there was 'no market' and the problems of filling and sealing a liquid crystal cell . . . were alleged to be 'insurmountable.'"[82] Joseph Castellano described one division manager opposed to the proposal because "if someone put the watch in his shoe at the beach, the display would turn clear at the high temperature," ignoring RCA chemists' proven ability to extend the operating temperature range of nematic materials for different applications.[83]

Freedman and his management team made convenient scapegoats for the holding pattern into which RCA's liquid crystal R & D operation entered by the end of the 1960s, but the tensions between Princeton and Raritan during this period were neither unprecedented nor unidirectional. Freedman's restrictive policies frustrated his staff and exacerbated disagreements with Somerville and the DSRC, but his priorities aligned closely with the 1969–1970 marketing report, which listed Heilmeier, Riddel, and Hiatt as principal contributors.[84] The rest of the Princeton team could not be cast as wholly blameless either, for tremors of discontent had started to emanate from the DSRC, foreshadowing a larger earthquake that would shatter the LCD development effort and the corporation as a whole.

"NIXIE TUBES ARE BETTER"

Freedman's efforts to bring RCA's LCD operation to heel agitated the previously strong institutional relationship between the DSRC and ECD liquid crystal groups. Defying the stereotypical characterization of RCA's central lab as an elitist country club indifferent to manufacturing issues, Heilmeier's team had worked closely with the Raritan

Figure 4.5. Richard Klein *(standing left)*, armed with a power drill and fake beard, hijacks the *Havana Special*—a mock-up airplane cockpit featuring dynamic scattering readouts. Also present are Sandor Caplan *(standing center)*, Ronald Friel *(standing right)*, Arthur Elsea Jr. *(seated left)*, and Louis Zanoni *(seated right)*. (Courtesy of Louis Zanoni.)

group ever since Klein's first visit in 1967.[85] Not only did Princeton staff members contribute materials and circuit designs to the LCD assembly line, they also partnered with Klein and Caplan to assemble new prototypes—such as a mock-up airplane cockpit—showcasing the utility of dynamic scattering (fig. 4.5).[86] Nonetheless, as the pilot plant inched closer to operation and the burden of responsibility for product development moved toward the operating divisions, some at ECD grew convinced that the DSRC team was no longer dedicated to the cause.

One could downplay Klein's recollections of the Princeton group's reluctance to help troubleshoot malfunctioning displays because that was not a sufficiently interesting technical problem as isolated incidents.[87] It is more difficult to ignore the shrinking number of people affiliated with the DSRC liquid crystal program. When the LCD went public in 1968, the group's red-covered reports listed sixteen people on the liquid crystal team's roster—not including consultants such as Richard Williams or Robert Lohman. By the first quarter of 1970, that number had been cut in half.[88] Much of this reduction stemmed from the decreased involvement of Bernard Lechner's CRL team following the termination of the matrix-addressed television project, but key liquid crystal researchers—personnel who had been with the group since its 1965 inception—had also begun to fall away. Some, including chemist Lucian Barton, transferred to other projects, such as the DSRC's new drive to develop a home video player.[89] Others, frustrated with the slow pace of LCD production, left RCA entirely to establish start-up firms based around the technology. By 1970, two of Heilmeier's earliest collaborators, Louis Zanoni and Joel Goldmacher, had joined the staff of one such company, Optel, which had been founded by former DSRC laser researcher Zoltan Kiss.[90]

More disconcerting than these reassignments or departures were those who believed that LCD technology had simply reached its limits. For many the bright promise of a dynamic scattering television had dimmed upon realizing the effort required to bring simple LCD applications into production. John van Raalte, another early member of Heilmeier's group and the first person to show a live TV broadcast on a liquid crystal screen, suggested this attitude pervaded all levels of the DSRC. "So we got very excited about liquid crystals," he noted. "We pursued it aggressively with a significant team of scientists and engineers for two, three years, at which time it became obvious to the management and to the technical direction that [the LCD] at least wasn't going to make it in ten years."[91]

Nothing captures this rising sense of skepticism within the DSRC better than a licensing report preserved in RCA's technical archive. The 1970 marketing study had recommended that no matter how the company proceeded with efforts to commercialize liquid crystals, it should offer other firms the chance to license its dynamic scattering patents.[92] Such a strategy aligned with RCA's traditional use of its intellectual property portfolio to control the radio and television industries.[93] To persuade domestic firms to purchase the rights to manufacture their own LCDs, RCA issued reports like the one in question here from September 1971, which summarized the science behind dynamic scattering and the engineering principles behind display production.

At least that was the document's original intent. Soon after its publication, however, an anonymous staff member, armed with a pair of scissors, a bottle of rubber cement, and a perverse sense of humor, decided to modify the report, scattering new words and images throughout the text. Many of these changes, such as the alteration of publicity photographs of project scientists or overlaying pornographic snapshots near otherwise staid images of laboratory equipment, had little to do with the document's ostensible subject matter. But the change in tone from the report's previously positive treatment of LCDs was evident from the very first page, where its innocuous title, *RCA Liquid Crystals*, was now amended with an ominous warning: "Be Suspicious."[94]

Where the text had once emphasized liquid crystals' amenability to large-scale manufacturing, the modified report cast LCD production as a messy, time-consuming process. A photograph of a technician laying out the coating patterns for a future display was replaced with an electronic hobbyist adjusting equipment on an overly crowded workbench. In a similar exaggeration, instead of a refrigerator-size glass sealing furnace, the report now showed an imposing multistory assembly line.[95] Perhaps the most telling of all was an image inserted on a page discussing the preparation of room-temperature liquid crystals. An added photograph showed a grinning scientist warming a kettle in front of an equation-filled blackboard. The man is smiling despite his bandaged fingers, presumably injured while preparing the beverage in his left hand. For all his apparent difficulties, a caption describes that process—and by extension, manufacturing LCDs—as "Simple and idiot-proof" (fig. 4.6).[96]

Nor were the revised report's critiques limited to factory procedures. The anonymous bricoleur went out of his way to discredit LCD applications. Mock-ups of possible products such as a dynamic

LIQUID CRYSTAL PREPARATION

Material preparation starts with purchased organic materials. The unique liquid crystal materials used in RCA devices are prepared by successive "reaction" and "purification" steps to provide 1.0 to 3.0 kg batch yields. All-glass, high-vacuum purification equipment is shown in Fig. 19. A portion of the materials process control laboratory is shown in Fig.20.

Figure 4.6. An anonymous critic modified this licensing report to highlight the disparity between the idealized version of LCD production captured in RCA marketing materials and the messy realities confronting personnel in Princeton and Raritan. (*RCA Liquid Crystals: Domestic Licensing* [Sept. 1971], 17. David Sarnoff Library Collection, courtesy of Hagley Museum and Library.)

scattering wristwatch and desktop calculator ("or mini-computer") were now framed as technological boondoggles, with the latter requiring a massive generator to operate properly.[97] Freedman and the RCA marketing staff had rejected those projects as impractical, but the changed report went further, denying any of the LCD's supposed benefits over other display technologies. The editor made his views on this matter quite clear, prefacing a list of those advantages with the caveat "NIXIE TUBES ARE BETTER."[98]

The Nixie tube was the first commercially successful electronic numeric indicator, a neon-filled gas discharge device that Princeton scientists hoped to replace with smaller, more rugged LCDs that operated at much lower power.[99] The assertion that they offered superior performance to dynamic scattering displays may not have been intended seriously. Like the rest of the report, it could be viewed as either a legitimate commentary on the LCD's shortcomings or a satirical caricature of liquid crystal skeptics. Either way, it illustrated an awareness of the technology's susceptibility to criticism from personnel at both the DSRC and the operating divisions. Such attacks, in combination with management's continued refusal to supply additional funding and the estrangement of the Princeton and Raritan groups left the LCD vulnerable at a particularly inopportune time.

RCA'S COMPUTER CRISIS

In January 1970, David Sarnoff resigned from RCA. Age and illness had sapped the General of his strength, and though he retained the title of honorary chairman, his son now held complete executive authority over company operations. Robert Sarnoff took advantage of his new role to accelerate his diversification strategy. While some of his earlier acquisitions, such as Alaska Communications Systems, reinforced RCA's traditional focus on electronics, more recent purchases, such as F. M. Stamper & Co. (later renamed Banquet Foods) and carpet manufacturer Coronet Industries, were harder to defend.[100] People joked that after buying these two firms and the Hertz rental car company, Robert Sarnoff's RCA no longer stood for "Radio Corporation of America" but "Rugs, Chickens, and Autos."[101]

The company's evolution into a conglomerate did not occur without protest. Robert Sarnoff faced condemnation at the February 1971 shareholders' meeting authorizing the Coronet purchase. He ended up securing the votes necessary to move forward with the deal but also fielded pointed questions about his decision to broaden RCA's business interests. "We have already from soup to nuts," one woman said, and then asked, "Tell me, mister, where is it going to end? You are going to build an empire, and look what happened to all the empires!"[102] Sarnoff replied that the sluggish national economy necessitated expansion into sectors offering greater growth opportunities.[103] He elaborated on this stance the following month in the firm's annual report, pointing out that expansion into service industries would grant RCA the freedom necessary to achieve its technical goals, notably "the establishment of a profitable computer business."[104]

RCA's electronic data-processing business had soldiered on in the years following John Burns's resignation, but as one manager commented later, there was still a "greater total effort in television from the engineering point of view than there was in the computer."[105] The pendulum started to swing in the opposite direction after the color TV boom, and by the time Robert Sarnoff assumed the presidency, RCA had achieved a modest commercial success with its Spectra 70 mainframes. Designed to be compatible with IBM's acclaimed System/360, the Spectra series heralded a new drive to compete against the nation's computing leader.[106] Sarnoff harbored no illusions of ousting IBM from its dominant place in the industry but believed it might be possible to leapfrog competitors such as Sperry Rand and Honeywell to secure the second-place spot. "In order to accomplish this goal," he vowed in a September 1970 speech, "RCA is prepared to commit whatever resources are necessary."[107]

Sarnoff repeated these claims over the course of the next year and launched an intensive sales campaign to publicize the firm's new "RCA series" of mainframes. While these systems offered only a modest increase in performance compared to the Spectra 70, company executives believed their lower costs would entice RCA's existing customer base along with a substantial number of IBM users.[108] This confidence was echoed in a 1970 promotional film, which declared the beginning of a "decade of difference" in the data-processing industry and reiterated Sarnoff's assertion that "RCA computers are going to be number two in the industry with 10% of the market by 1975."[109]

Unfortunately, the company's rhetoric exceeded the capacity of its Computer Systems Division (CSD) to keep pace with IBM. CSD reported significant annual losses in 1969 and 1970, and despite promises to shareholders that a crossover to profitability was imminent, the RCA series was another financial disappointment.[110] Internal audits during the first half of 1971 revealed that Sarnoff's optimistic predictions had been based on faulty bookkeeping.[111] A revised estimate suggested that turning RCA's computer business around would require a billion-dollar capital investment by 1976, half of which would go directly to CSD. "I can think of two dozen things I would rather spend $500 million on," one member of the company's board of directors quipped upon learning this figure.[112]

Sarnoff, who had staked his reputation on the success of RCA's computer business much like his father had on color television, ignored these findings and resisted pressure to abandon his crusade. Only in the summer of 1971, after America entered into a recession and further evidence accumulated confirming CSD's financial instability, did he concede "that in view of everything we could not afford the price of staying in."[113] On September 16, he received authorization from the board of directors to withdraw from the general purpose computer market. Shortly thereafter, the company sold CSD to the Univac Division of Sperry Rand for $127 million. It also set up a pretax reserve of $490 million ($250 million posttax) to cover prospective losses in connection with the sale.[114] As the *Wall Street Journal* reported, at the time the write-off was "probably the largest in corporate history, exceeding even Ford's Edsel."[115]

Despite rumors of a power struggle within the corridors of 30 Rockefeller Plaza, Sarnoff remained chairman following the computer divestiture. He promptly imposed a freeze on acquisitions and implemented what the company described as the "most stringent cost-reduction program in RCA history," one aimed "at all corporate and divisional levels, affecting administrative and operating units alike."[116] Factories closed across the country, and the DSRC, where nearly half the staff had been drafted into computer-related work, took an immediate 10 percent budget cut and a corresponding 6–7 percent staff reduction.[117] All projects were subject to immediate review, placing those with questionable commercial potential, such as the LCD, in a precarious situation. Within a year, the already shrinking Princeton liquid crystal team dropped to a paltry half dozen members as personnel were let go or transferred to other groups.[118]

The cuts at the ECD pilot plant in Raritan were even more extreme. According to Klein, "In 1971, when they canned the computer division, our group was, I think, thirty-one people . . . and they cut the group back to eight."[119] The survivors, who transferred back to the recently renamed Solid State Division (SSD) in Somerville, included Freedman and Lawrence, the two men whose policies had most isolated the liquid crystal operation from the rest of the corporation.[120] Neither Klein nor Caplan retained their positions, but they soon obtained new jobs with Ashley-Butler, the advertising firm that sponsored the development of their point-of-purchase displays.[121]

Liquid crystal researchers constituted only a tiny fraction of the estimated 13,000 people who lost their jobs because of RCA's withdrawal from the computer industry, but this purge was the culmination of policies that had consistently undermined their efforts to move the LCD toward commercial production.[122] Robert Sarnoff's acquisition-fueled venture into electronic data processing diverted resources away from consumer electronics products that required long-term investment. This lack of support and deteriorating ties between the DSRC and the rest of the corporation aggravated preexisting anxiety about the future of dynamic scattering displays and spurred the exodus of LCD personnel before and after the 1971 layoffs.

LIQUID CRYSTALS AND THE LIMITS OF THE LINEAR MODEL

RCA did little to staunch the outflow of liquid crystal researchers, with one notable exception. Since 1964, George Heilmeier had been the LCD's strongest advocate at the DSRC, but the infighting between Princeton and Raritan had sapped his passion for the work.[123] He considered leaving the company, like his friends Joel Goldmacher and Louis Zanoni. The corporation, fearful of losing one of its most prominent engineers, had already named Heilmeier head of solid-state device research at the DSRC and now offered him a new title, head of device concepts research, to entice him into staying put.[124] When that failed to prevent him from broaching the possibility of forming his own firm, management replaced its carrots with sticks. Heilmeier recalled being warned that "if I left RCA to start my own company, it would be the biggest mistake of my life because they would sue the hell out of me."[125]

A White House fellowship ultimately supplied Heilmeier with the means to escape Princeton in 1970 and begin a new career in government service. At the time he downplayed any frustration he may have felt toward his superiors. When asked in an interview shortly after arriving in Washington what prompted his decision to leave RCA, he did not comment on the apparent bungling of the LCD, responding instead that he "really wanted the opportunity to see how people at very high levels operate and how they handle their daily problems."[126] For its part, the corporation treated Heilmeier's career change as temporary, stating that he was "on leave from RCA Laboratories" well after the conclusion of his one-year assignment as special assistant to secretary of defense Melvin Laird.[127] But while Heilmeier's name remained linked with RCA and liquid crystals for the rest of his life, his absence from the DSRC was permanent.

Liquid crystal display research would survive at RCA's laboratories and factories without George Heilmeier, but the circumstances of his departure cast doubts on the DSRC group's ability to guide the commercialization of dynamic scattering displays and the linear model as a whole. The confidence that had encouraged their scientific investigations had yielded to disillusionment, and the once limitless possibilities of liquid crystals were now constrained to a handful of marginal product lines. The barriers between research and production, which had vanished briefly during Princeton's collaboration with Somerville, had arisen again, complicating any future attempts at technology transfer.[128]

Reflecting on his experiences in 1976, Heilmeier suggested that the situation might have evolved differently had the Princeton liquid crystal team been responsible not just for the development of the

LCD but also its commercialization.[129] After all, the scientists and engineers on his team "were the entrepreneurs; the ones who saw opportunities, not problems; the ones who had no vested interest."[130] Such individuals, he argued, could not thrive within the confines of a large established corporation like RCA. Instead, Heilmeier believed that new technologies were more likely to emerge in smaller organizations without any obligation to maintain the status quo. "If one subscribes to this theory," he concluded, "it is not surprising that the polaroid [*sic*] process was not nurtured by the largest photographic company in the world, that most vacuum tube companies did not make it in the transistor business, and that Xerox office copiers were not pioneered by the giants in the office equipment business."[131]

Essentially, Heilmeier viewed the LCD as what innovation expert Clayton Christensen would later refer to as a "disruptive technology," one whose attributes did not sustain or enhance existing RCA product lines, at least in the near term. Disruptive technologies often compare poorly to earlier products when judged using conventional performance metrics even though they are "typically cheaper, simpler, smaller, and frequently, more convenient to use."[132] They also open up new markets, but these are generally too small to support the growth needs of well-established firms.[133]

All of these criteria are applicable to the LCD project at RCA and reinforce the conclusion held by Heilmeier and many of his colleagues that the major obstacle to the commercialization of liquid crystals was RCA's management.[134] For all of their advantages, the lightweight, low-power, dynamic scattering displays that the DSRC and ECD groups assembled could not match the picture quality of existing televisions and would not persuade the company to abandon its long-standing investment in the CRT. Jack Riddel's development contracts confirmed that there were audiences interested in dynamic scattering, but the benefits of pursuing them seemed insignificant compared with the potential profits to be found in digital computing.

Managerial misconduct undoubtedly compounded the challenges associated with transforming the DSRC's flat-panel prototypes into commercial displays. Yet as much as one might blame Robert Sarnoff's strategic missteps or Norman Freedman's domineering oversight of the ECD pilot line for RCA's poor handling of the LCD, it is important to recall that they did not act in complete isolation from the company's technical staff. Upon encountering an unfamiliar technology, RCA executives reached out to the DSRC and ECD to evaluate the commercial potential of dynamic scattering. Recall that Heilmeier, Riddel, and Hiatt all contributed to the 1970 marketing report, which concluded that the LCD was mostly irrelevant to the corporation's business. Frustrating as Freedman's policies may have been to personnel in Raritan, his decision to ignore wristwatches and concentrate on advertising displays, car mirrors, and numeric counters corresponded exactly with those findings.

In other words, the determination that the liquid crystal display was a disruptive technology occurred, at least in part, as a result of discussions with members of RCA's technical staff. Their growing experimental familiarity with the material logic of liquid crystals led them to tamp down the high expectations set forth at the 1968 press conference. The recognition that matrix-addressed LCD televisions would not be arriving in the near future drove some researchers to explore simpler devices with renewed zeal while others began to see liquid crystals as a dead end.

The conservatism with which both groups regarded the LCD had important consequences. It seeped into the technical memorandums directed toward managers at the DSRC and beyond, discouraging them from funneling more resources into the ECD operation. The latter group's subsequent reliance on contract funding and the widening schism between Princeton and Raritan provoked additional fears about the project's future, leading to increasingly pessimistic forecasts. The culmination of this vicious cycle was the partial dissolution of the DSRC liquid crystal team and, in the aftermath of the computer crisis, dramatic cutbacks at ECD.

Heilmeier and the other LCD personnel who left RCA had legitimate reasons to be disappointed with the company's half-hearted commercialization of dynamic scattering displays. Where the formation of the liquid crystal research team confirmed that scientists and engineers could exert influence within a corporate laboratory, the technology transfer process showed that there were limits to that authority. They were not, however, completely powerless. The ultimate outcome was not what they originally hoped, but through technical reports, marketing studies, and face-to-face conversations, the DSRC research group and their ECD allies did influence how RCA executives approached the LCD. Those liquid crystal researchers who stayed at the DSRC after Heilmeier's departure would remain engaged in discussions of flat-panel development even as the presence of new competitors and display technologies left the LCD's future at RCA in a state of profound uncertainty.

5 THE CHANGING OF THE GUARD, 1969–1976

Throughout the summer of 1971, a general lay dying in Manhattan. The 1968 shingles diagnosis that prompted David Sarnoff to retreat from the boardroom had given way to a series of mastoid infections that ravaged his body. Despite multiple surgeries to reverse his condition, the once indomitable executive withered to a haggard shell, bedridden on the top floor of his six-story townhouse. Early on Sarnoff arranged for regular briefings from RCA staff members, but his weakening health soon made it difficult to see, hear, or speak.[1]

This isolation initially prevented him from discovering that his corporation was in peril. Friends and family tried to block Sarnoff from any news related to the failure of RCA's Computer Systems Division. To their chagrin, the broadcast networks he had championed in his youth undermined their efforts. In September 1971, a nurse commented on the firm's withdrawal from the computer business after hearing about it on the radio. Mustering up the strength to speak, Sarnoff referred to the news as a "terrible tragedy," though he could take solace that, for the time being, his son Robert would remain chairman of RCA.[2]

Two months later, on December 12, NBC interrupted a *Meet the Press* interview with Secretary of Agriculture Earl Butz to announce that David Sarnoff had passed away.[3] Funeral services were held the following Wednesday at Temple Emanu-El, the largest synagogue in the United States. New York Governor Nelson Rockefeller, whose father partnered with Sarnoff to construct the Radio City complex that housed RCA's headquarters, delivered the eulogy before a crowd of seven hundred people. That day there was no mention of Sarnoff's feuds with the FCC and Justice Department or his clashes with other entrepreneurs over the future of America's airwaves.[4] Instead, Rockefeller praised the General as a visionary, citing "his capacity to look at the same things others were looking at—but to see far more. Others looked at radio and saw a gadget. David Sarnoff looked at radio and saw a household possession capable of enriching the lives of millions."[5]

Among those most directly affected by their association with David Sarnoff were the scientists and engineers at his beloved Princeton laboratories. As they gathered in the DSRC's auditorium to watch Rockefeller's speech, many wondered what would become of RCA in Sarnoff's absence. "Many of us considered him a personal friend," an internal DSRC newsletter noted. "He enjoyed coming to Princeton and talking to the research staff. On more than one occasion the vision he perceived from such talks led to significant developments in the fields of electronics and communications."[6]

Certainly the LCD—and other RCA flat-panel projects—had benefitted from Sarnoff's patronage. The General's desire to create "the television of tomorrow," which would hang on the wall like a painting, had inspired two decades of research. In recent years, however, the campus-like atmosphere of the DSRC had become more entangled in applications work underway at the operating divisions. Now the disintegration of Robert Sarnoff's computing initiative left RCA without a clear R & D strategy.

William Webster and the rest of the DSRC's leadership rushed to convince the technical staff that the conclusion of "our forced draft effort on behalf of Computer Systems" would ultimately benefit the laboratories.[7] Even with major staff and budget cuts, they promised that "our program moving into 1972 is better balanced and, we hope, more responsive to the entire spectrum of RCA research needs."[8]

Foremost among those needs was a sustained effort to create a home video player, which the DSRC framed as a logical extension of RCA's traditional focus on home entertainment.[9]

Over time, this "SelectaVision" project held increasing sway over the DSRC's research agenda as the company passed over a preliminary laser-based approach using hologram-embossed tape in favor of the VideoDisc system, which stored picture information in the grooves of a conductive vinyl record.[10] The prioritization of VideoDisc eclipsed RCA's attempts to develop applications for dynamic scattering LCDs and build a flat-panel television, effectively decoupling those two goals. Liquid crystals had their place in simple products such as numeric indicators, but few at RCA dared to fantasize that they would ever replace the CRT or restore the company's now sullied reputation.

At first glance, the declining fortunes of RCA's liquid crystals between 1971 and 1976 might seem unavoidable, the preordained fate of a disruptive technology within a large corporation. Clayton Christensen suggests such firms find themselves at a disadvantage when approaching products that lack a clearly defined market and whose limited early profits discourage expanded investment.[11] In contrast, the smaller scale and greater flexibility of start-up firms leaves them better prepared to embrace disruption.[12] An intuitive understanding of this dynamic, which had played out numerous times in the semiconductor industry during the 1960s, informed the decisions of several members of RCA's technical staff to leave the DSRC and establish their own LCD businesses.[13]

Examining the earliest and most influential of these organizations, Optel, alongside its parent company, allows us to consider the challenges confronting both types of firm as each sought to commercialize the LCD. RCA possessed the resources to become a major player in this new industry. Its operating divisions started fabricating dynamic scattering displays, but company managers, still reeling from the fallout of the computer collapse, consigned them to a handful of minor product lines and reduced liquid crystal research funding at the DSRC. Optel was a much smaller operation, yet its proportionally greater commitment to push the limits of LCD technology allowed it to create the world's first digital wristwatch with a liquid crystal readout just a few months after RCA dismissed the idea as impractical.

When all was said and done, neither RCA nor Optel established itself as a lasting presence in the global flat-panel display market. The reasons for their respective downfalls varied sharply, but each firm's experiences in the early 1970s reveal that the LCD's status as a disruptive technology, while obvious in hindsight, hinged on shifting understandings of the material logic of liquid crystals. No matter how large the company, the scientists and engineers who engaged most directly with these LCD projects continued to influence which ones received further attention and which would be left for others to pursue.

ON THE ORIGINS OF OPTEL

Both the research team behind the development of the LCD and the first spin-off to unlock its commercial potential emerged from the DSRC's quantum electronics group. Established in October 1960, shortly after Theodore Maiman demonstrated a functional ruby laser at Hughes Aircraft, the group possessed a broad mandate "to do research on stimulated-emission phenomena and devices."[14] Under the leadership of Henry Lewis, researchers concentrated on gaining greater familiarity with existing laser equipment and devising new ways to produce, detect, and modulate laser light.[15] The last of these objectives had captured George Heilmeier's attention and eventually led him to explore the electro-optic properties of liquid crystals.

In the midst of these investigations, Heilmeier occasionally crossed paths with another laser researcher named Zoltan Kiss. After being arrested for protesting the Communist occupation of his native Hungary, Kiss had escaped to Canada and obtained a doctorate in physics from the University of Toronto in 1959. He then signed up for a yearlong postdoctoral fellowship in Oxford, where he became fascinated with lasers. "That's what I really wanted to do," he recalled. "But at that time, Oxford didn't have any lasers yet."[16] Luckily, Kiss had already secured a position on RCA's technical staff before leaving for

Europe, and after returning to Princeton, he was assigned to Lewis's group.

Kiss partnered with theoretician Donald McClure, phosphor chemist Neil Yocom, and a few technicians to seek out new compounds capable of generating laser light.[17] On McClure's recommendation, they began by reaching out to Simon Larach's materials research group for help growing crystals of calcium fluoride, which scientists at IBM had fashioned into a laser at the end of 1960.[18] Instead of following IBM's example and doping their crystals with uranium, Kiss and his colleagues mixed in small amounts of rare earth elements into the calcium fluoride. They determined that samples containing divalent rare earth atoms (i.e., those lacking two electrons in their outermost orbits) absorbed a greater amount of incident light and emitted more energy than other compounds, suggesting a possible avenue to a more efficient laser.[19]

Previous lasers had relied on bright artificial lamps to bombard crystalline materials and induce them into emitting light whose photons shared the same energy and frequency—a process called "pumping." This coherent beam remained focused over long distances, which RCA scientists believed might revolutionize telecommunications if they could reduce the amount of power involved.[20] Kiss and his team made a promising step toward that goal in 1962 when they assembled an infrared laser—using calcium fluoride laced with dysprosium—that could be pumped using sunlight (fig. 5.1).[21] "That was sort of my baby," Kiss remembered, "because you couldn't do it with anything else. . . . You could only do it with this divalent rare earth."[22]

RCA's new laser received a lot of media coverage. The *New York Times* published James Hillier's descriptions of "sun-powered lasers on future space satellites" that could be used for either communications or geodetic measurements.[23] What Hillier did not mention was that before any of those predictions could come true, Kiss's invention would have to overcome a serious flaw. Like fading paint on the outside of a house, prolonged exposure to the sunlight used to pump the laser bleached the calcium fluoride crystal, ruining its performance.[24]

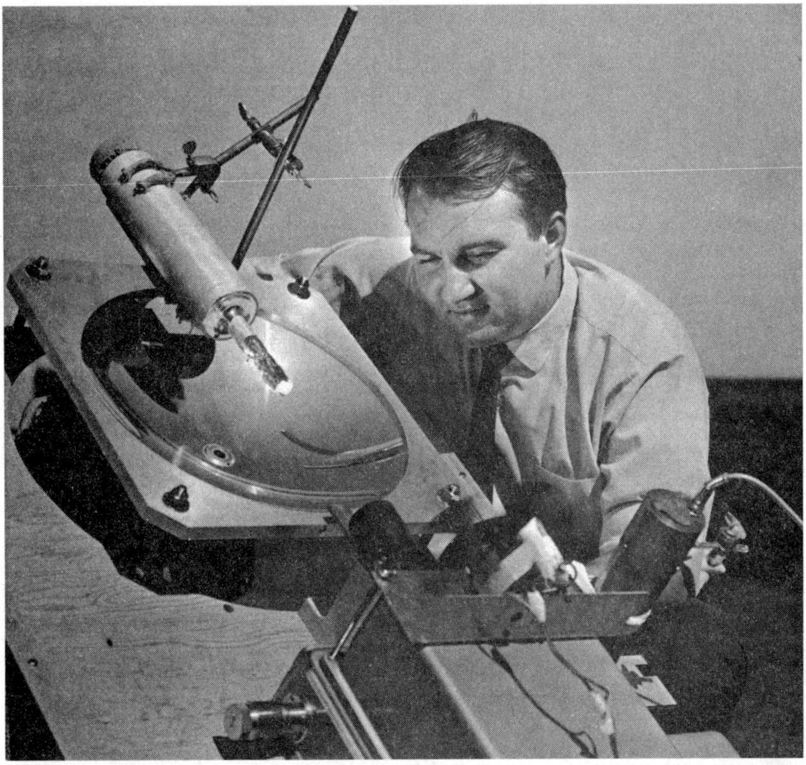

Figure 5.1. Zoltan Kiss operating his sun-pumped laser on the roof of the DSRC. (*RCA Radiations* [Oct.–Nov. 1962], cover. David Sarnoff Library Collection, courtesy of Hagley Museum and Library.)

As a preventative measure, Kiss started "double doping" his crystals, adding tiny amounts of two different rare earths. This procedure eliminated the bleaching effect. It also resulted in the synthesis of several new materials exhibiting an unexpected property. "I would start out with a red crystal," Kiss explained. "I would shine light on it, and a few minutes later, now it's blue. What the hell is this?"[25] Even more perplexing, he was able to produce materials that changed color—from blue to red, for example—when struck with one wavelength of light, but reverted back to their original color when hit with light of a different wavelength.

Kiss imagined combining these new "photochromic" materials

with lasers to create a new information storage and recording medium.[26] The discovery that one could also trigger a color change in certain compounds using an electron beam, like those found in CRTs, also hinted at a possible display application. Unlike conventional picture tubes, where images had to be refreshed constantly, the pictures painted on a "cathodochromic" screen (named for its reliance on cathode rays, not light, to change color) would stay in place until they were erased.[27]

Kiss's transition from laser research to photochromic and cathodochromic materials paralleled the broadening mission of the quantum electronics group.[28] It also positioned him to lead the group after the newly promoted William Webster selected Henry Lewis to run the DSRC's Materials Research Laboratory in 1968.[29] Kiss adapted quickly to his new managerial responsibilities and got along with his team well enough that they frequently stayed after hours to work on their latest technical breakthrough. Indeed, his tendency to drop into researchers' offices asking "Breakthrough? Breakthrough?" made the question into something of a personal catchphrase.[30]

Management agreed with Kiss, but his professional success did little to stifle a growing sense of restlessness. He had sought out a career in industrial research to create products that would improve people's lives, but no matter how many remarkable applications he contrived for his laser research, "the quantum electronics group was considered more as a basic research group."[31] As he told *Fortune* in 1973, RCA was ignoring potentially significant inventions "because their sales potential did not seem big enough to division heads who were accustomed to thinking in terms of $100-million and $200-million markets."[32] In short, company decision makers viewed Kiss's lasers and color-changing crystals as disruptive technologies. If he wanted to commercialize them, it would not be under RCA's banner.

Kiss pondered his options during a sabbatical at RCA's Zurich laboratory and finally approached the DSRC's management with an idea. Inspired by the popularity of start-up firms in the region that was becoming known as Silicon Valley, he would establish his own company to capitalize on the quantum electronic technologies RCA cast aside.[33] Since RCA retained the relevant patents for his work on photochromic and cathodochromic materials, he would need to pay a licensing fee. Of course, RCA was quite used to such arrangements, and this proposal seemed particularly advantageous because the company had no plans to utilize Kiss's crystals and did not expect to lose any customers to his new business.[34]

After listening to Kiss's arguments in the summer of 1969, Webster and Lewis granted him their blessing to begin laying the groundwork for his new company, initially called Quantel because of its focus on quantum electronics. The firm started operations with that name in December 1969 but abandoned it after learning that another business had previously registered the brand. By the end of 1970, its staff settled on a new name referencing the start-up's location at the intersection of optics and electronics. They called it Optel.[35]

A SPIN-OFF DIVIDED

In November 1969, the DSRC quantum electronics group arranged a farewell party for Zoltan Kiss. During these festivities, they gave him an album filled with snapshots from his time in Princeton as well as a remarkable caricature drawn by one of his technicians, Robert Quinn.[36] The latter, titled "Zoltan's Last Supper," replaced the figures in Leonardo da Vinci's mural with Kiss and his research team. Several of the "apostles" are shown making jokes at Kiss's expense. "Hallelujah," one remarks, "we won't have to hear that dreaded word, 'BREAKTHROUGH', again." Another jeers that "'Zolly' thinks that having your own business is a bed of roses—he'll find out!!" All the while, a haloed Kiss gazes outward with a somber expression, his perspective captured in a cartoon thought bubble: "Oh Lord—Deliver me from these imbeciles!" (fig. 5.2).[37]

Contrary to this satirical inner monologue, Kiss truly respected his RCA colleagues, and the first people he recruited to join his new start-up came from within the company. Recognizing the need for personnel with manufacturing experience, Kiss reached out to

Figure 5.2. "Zoltan's Last Supper," a parody of the Renaissance masterpiece, drawn by one of Kiss's technicians, Robert Quinn (aka "Leonardo DiQuinci"), before the physicist's departure to form Optel. (Courtesy of Zoltan Kiss and Robert Quinn.)

collaborators at the company's operating divisions, most notably the Aerospace Systems Division (ASD) in Burlington, Massachusetts. ASD oversaw most of RCA's defense-related laser projects, including communications systems, satellite trackers, and range-finding equipment.[38] Kiss was an occasional visitor to Burlington and eventually persuaded physicist Edward Kornstein and electrical engineer Nunzio "Tony" Luce to join his new enterprise.[39] He also enlisted allies closer to home, recruiting spectroscopist Douglas Bosomworth from the DSRC quantum electronics group as well as two former members of Heilmeier's liquid crystal team, Joel Goldmacher and Louis Zanoni.

In some ways, Zanoni and Goldmacher were unlikely Quantel employees. After all, Kiss's work had focused on solid, inorganic materials, not the mesomorphic, organic compounds that exhibited dynamic scattering. The situation makes more sense when one recalls that both Kiss and Heilmeier reported to Henry Lewis and attended regular group meetings to discuss their research. As a result, Kiss became aware of Heilmeier's LCDs before RCA presented them to the public. He came to believe that company higher-ups were treating these liquid crystals similarly to his solid ones, again ignoring the possibility of exciting new display technologies to concentrate on the existing television market. "I was surprised about this," Kiss noted, "and that was one of the main reasons when we started Optel . . . that we decided to sort of look at liquid crystals as a potential commercial activity."[40]

There was also another, more pragmatic reason for Kiss to latch on to liquid crystals during Quantel's early days. Although he harbored grand ambitions for his cathodochromic and photochromic materials, he had not yet incorporated them into a product that would captivate investors. Dynamic scattering prototypes, on the other hand, had been around for several years, most of them built by Heilmeier's technician—and recent Quantel recruit—Louis Zanoni. During meetings with New York venture capitalists, Kiss discovered that even the simplest LCDs could lend credibility to his technical presentations. On one occasion, after describing his firm's interest in lasers, photochromism, cathodochromism, and liquid crystals, an audience member replied that "all were interesting, but we can't really understand too well any of these. Can you demonstrate something?" Kiss responded by deferring to Zanoni, who presented a dynamic scattering window "maybe a half a square centimeter. You throw a switch, and it's shiny white. Throw it in the other direction, and it's transparent." It was a simple demonstration, but enough to win over the crowd.[41]

As a result of Kiss's salesmanship and Zanoni's demonstrations, by the fall of 1969, Quantel had accumulated approximately $1 million in private capital and development contracts, enough to purchase a building to serve as a base of operations.[42] Promotional materials later referred to this space, located a few miles north of the DSRC on US Route 1 as "a converted Princeton garage," though in subsequent interviews staff members disclosed that it was originally a pharmaceutical research facility.[43] The building had room for a fully stocked materials science laboratory, including crystal growing furnaces and vacuum deposition equipment.[44] Unlike at RCA or its peers (GE, IBM, AT&T, etc.), whose adherence to the linear model had driven them to physically isolate their research activities from their factories, this ten thousand square-foot space would combine both functions under one roof.[45] This mirrored the approach popularized by Intel, whose founders felt that separating R & D from production complicated the technology-transfer process.[46]

Everything appeared ready for Quantel to commence operations in early 1970, but almost immediately debates erupted over how the company should proceed. In marketing materials, Kiss framed Quantel as a firm that would commercialize cathodochromic, photochromic, and liquid crystal materials, but internally, he expressed a clear preference for cathodochromics. Not only could they be readily substituted into CRT-based displays, they also seemed more amenable to mass production than photochromic memory systems because of the latter's reliance on lasers.[47] As for liquid crystals, Zanoni's impression upon joining the staff in 1970 was that Quantel viewed dynamic scattering as a "backup technology."[48] Kiss could not wholly ignore the LCDs that had convinced so many to support his company, but he apportioned far fewer resources to their development. "Their primary technology then was cathodochromic displays," Zanoni recalled. "I guess ninety percent of the company were working on that."[49]

The division of labor was not quite that stark, but for much of 1970 the majority of Quantel's staff set themselves to work on the Reflicon, a cathodochromic display meant for use in computers and facsimile systems. Quantel engineers also embedded the Reflicon into a data terminal that could display text or images and print them using a built-in photocopier.[50] At the same time, the company's liquid crystal group, consisting of just Zanoni and Brazilian technician Amilcar Guimaraes, focused on building sample dynamic scattering displays.[51] Later, Kiss assigned Joel Goldmacher and Tony Luce to Zanoni's LCD team, but for the rest of Quantel's twenty-person staff, the key objective was preparing the Reflicon for its November 1970 debut at the Fall Joint Computer Conference in Houston.[52]

Everyone at the company recognized the significance of the Texas trip. Houston was where Kiss would introduce the firm's first product line as well as its new name—Optel. The Princeton start-up was about to step on to the national stage, and while the Reflicon would receive the most attention, the liquid crystal group also wanted to participate. “So I got together with Lou [Zanoni] and we sat down, and Joel Goldmacher, and we said let’s come up with a whole slew of displays,” Tony Luce told interviewers later.[53] Together, they assembled a collection of alphanumeric LCD readouts of various sizes. “And we took all that stuff, we put it on a big board and we went to Texas with that display,” Luce continued. “And we went down to the show, we took the whole show away from the cathodochromics. It was really very spectacular.”[54]

In actuality, Optel’s LCDs did not totally overshadow its cathodochromic displays. In their coverage of the Houston conference, *Business Week* and *Electronic News* reported on both technologies, with the former noting that each promised “to make electronic information easier to read in light that washes out normal displays.”[55] Still, to some within the company, the fact that Optel’s liquid crystals had generated enough interest to be mentioned even in passing suggested that the company should reexamine its priorities. Edward Kornstein, who had been recruited to work on cathodochromics and wrote some of Optel’s earliest business plans, now believed that LCDs offered a greater possibility of long-term growth.[56] Kiss meanwhile remained committed to the status quo, with liquid crystals occupying a subordinate position. “And I think in his mind he could see that the cathodochromic technology wasn’t going to be going anywhere,” Tony Luce explained. “But you know, he’d drag things out, like most people do when it’s their baby. You’re going to drag it out as long as you can, even though it’s dead.”[57]

OPTEL PICKS ITS PATH

The tensions over Optel’s future came to a head at the end of November 1970. A few days after the Reflicon demonstration in Houston, Kiss left on a previously scheduled European fundraising trip. His original itinerary called for him to return to the United States on Sunday, November 29, but his meetings ended earlier than expected and he decided to catch an early flight home. Upon arrival in New Jersey on Saturday night, he discovered a telegram waiting for him.

Apparently Kornstein and two other members of Optel's board of directors had grown upset with Kiss's management and scheduled a meeting in New York to remove him from office on Sunday, when he was supposed to be in transit.[58]

The scene was set for a dramatic showdown. Kiss startled his opponents by appearing in the boardroom for the Sunday meeting with several of his supporters, including members of the company's liquid crystal group. Although Kiss had bestowed only limited support for their projects, they still felt a strong sense of loyalty to Optel's founder. Luce expressed disdain for the defectors' tactics. "I didn't think that people should just go ahead and do something behind someone's back," he noted. "Hold a special board meeting and throw the president out."[59] To Zanoni, it was also unclear whether Kiss's self-styled successors would be any better than the man they were replacing. "Within the group there was a choice of do you want Zoltan to be chairman or these other guys to be chairman," he observed. "And Zoltan . . . I had worked with Zoltan. . . . So we stuck with him, and of course he won over eventually."[60]

Kiss did manage to thwart his foes in the ensuing negotiations, which stretched into the early hours of Monday morning, but not without paying a high cost. He kept his position as Optel's chairman and forced Kornstein and his allies to step down from the board, but in exchange for these concessions, he had to sign away a quarter of his stock. He no longer retained majority control over the company. In addition, the rebellion led Kiss to reevaluate how Optel allocated its R & D budget. If given the chance to do it all again, he told a reporter a few years later, "I would have spent only 60 percent of what we did on cathodochromics."[61] The LCD group had stood by Kiss in his hour of need, and he would not forget them now.

This change of heart could not have been better timed, because in December 1970, Tony Luce completed the project that would put his company on the map. When Optel discussed its liquid crystal research in Houston, it emphasized product lines that replicated work already done at RCA, including numeric readouts and cockpit instrumentation. What the public did not know was that the firm was also working on a wristwatch with a dynamic scattering display.

The origins of the project remain ambiguous. Kiss claimed later that he set the technical direction at Optel and that "the basic elements, how are we going to do it, that came from me . . . What is the most highest volume thing that this can use? Digital watches came from me."[62] Luce rejected this claim outright. When asked whether the LCD watch was Kiss's idea, his response was blunt: "No, absolutely not. The watch came out actually almost by accident, I'll be very honest with you."[63] Zanoni provided an answer that lay somewhere in the middle. Like Kiss, he stressed the importance of economic considerations, since "the display had to be small and there had to be a market that they would pay a higher price for."[64] According to him, the move from those general criteria to a wristwatch resulted from discussions with Optel's marketing staff, not Kiss. It is also possible that Zanoni and Goldmacher, who were aware of RCA researchers' interest in LCD wristwatches, carried the idea with them from the DSRC.

In any case, as early as June 1970, the liquid crystal group had drafted a technical proposal for the Bulova Watch Company outlining a "best effort program to develop a prototype of a liquid crystal display and its driving electronics for an electronic wrist watch."[65] Bulova had already advanced the transition from mechanical to electronic timekeeping with its Accutron model, a watch that used a battery-powered tuning fork instead of a balance wheel to tick off the seconds.[66] Now Optel (still known as Quantel at that point) proposed a partnership to build a fully electronic watch with a dynamic scattering readout.

Just a few months earlier, this idea would have sounded outlandish, but in May 1970 the Hamilton Watch Company announced that it had created the world's first digital wristwatch: the Pulsar. The Pulsar replaced the Accutron's tuning fork with a vibrating quartz crystal—an approach pioneered by Japanese watchmaker Seiko—and its mechanical dial with a set of bright red light-emitting diodes. Because they generated their own light, these LEDs required more power to operate than an LCD. As a result, the Pulsar only displayed the time when users pressed a small button on its side. The effect was dramatic enough to merit a cameo in the next James Bond movie (*Live and Let Die*), but it also served as a constant reminder of the watch's short battery life.[67] The Quantel proposal explicitly brought up this problem, commenting that the LED's "much greater power requirements (more than 1000 times greater) and its low brightness cannot compete with liquid crystals."[68]

Bulova's executives greeted Quantel's proposal with skepticism. "They had the Accutron, and that's it," Tony Luce recollected. "That's the watch, and all this LCD is really, quartz is not the way to go. . . . But their research people were very forward looking also. They kind of let us go ahead with this program, which we did."[69] With Bulova's backing, the Quantel team set to work in the summer of 1970, with Goldmacher, the chemist, synthesizing batches of room-temperature liquid crystals that could be inserted into Zanoni's numeric readouts.

For his part, Luce wrestled with the task of designing the circuitry for an LCD watch. "Because LCDs had a lot of problems then," he acknowledged. "We were doing it with dynamic scattering, and you had to drive it at fifteen volts, and where are you going to get a chip to drive it at fifteen volts. There was a lot [of] little things that had to really be done."[70] Never one to turn away from a technical challenge, Luce took each problem in stride. Looking back, Zanoni extolled his friend's work ethic. Luce, he said, "could work twenty-four hours straight, nonstop. . . . I think in less than a week he did design that watch circuit."[71] The resulting breadboard (fig. 5.3) possessed an architecture similar to the Pulsar, including circuits to count the vibrations of a quartz crystal, divide them into seconds, minutes, and hours, and activate the appropriate digits on a liquid crystal readout. The LCD's lower power requirements meant that, unlike the Pulsar, the time could be shown continuously without rapidly draining the battery. To confirm the watch was working, Luce also added a feature that would become almost universally adopted in digital watches and clocks: a colon that blinked once every second.[72]

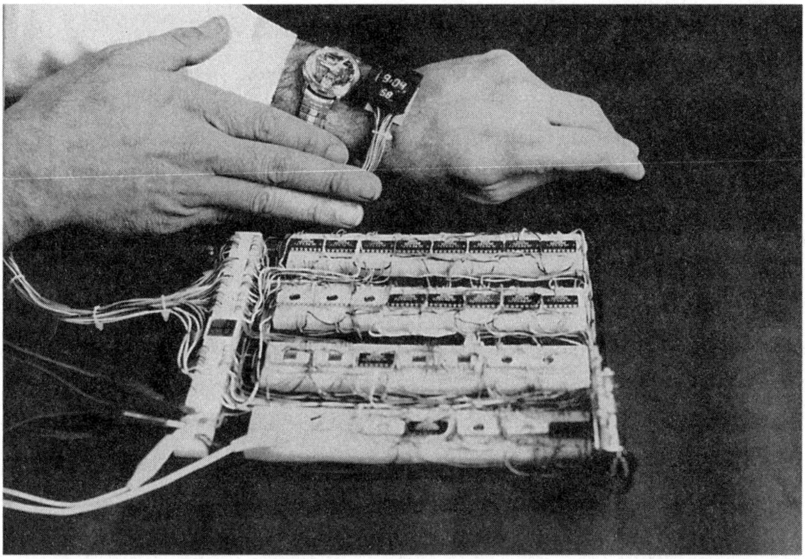

Figure 5.3. Nunzio Luce tests out the LCD watch circuit he assembled at Optel in 1970. Optel engineers soon compressed this breadboard on to a pair of integrated circuits small enough to fit on a person's wrist. (Courtesy of Louis Zanoni.)

Luce presented his creation to Kiss, and as he put it, "the dollar signs lit up in Zoltan's head."[73] Although he continued to call attention to hope for his cathodochromic displays during the latter half of 1970, he seized on the LCD wristwatch as both a product line and fundraising tool. As the liquid crystal team contacted local integrated circuit manufacturers for assistance reducing Luce's breadboard to something that could fit on someone's forearm, Kiss reached out to watch companies intrigued by the idea of digital timekeeping. He found a very receptive audience in Switzerland, where firms that had taken pride in their mechanical craftsmanship now found themselves at a loss for electronic expertise.[74] Kiss's frequent visits to Europe opened up new sources of operating capital but stirred up discontent within the firm. It was following one such meeting with Swiss watchmakers that Kornstein and his colleagues attempted their coup.

Whatever complaints Kiss's travels inspired disappeared in the weeks after he consolidated his authority, especially after the delivery of Luce's watch circuits. Optel partnered with Solid State Scientific, a small semiconductor firm in the Philadelphia suburbs, to condense nearly 1,300 transistors on to a pair of tiny silicon chips.[75] Like the LCD, these CMOS (complementary metal-oxide-semiconductor) circuits traced their origins to the DSRC.[76] While RCA personnel suspected that their low operating power might be useful in wristwatches, the firm never invested in the idea, preferring to supply CMOS chips to outside firms—including Hamilton, which used them in the Pulsar.[77] Only after the Optel team completed their prototype watch would these two RCA technologies—CMOS and the LCD—be united inside a functional timepiece.

"Functional" was a somewhat generous description of that first demonstration model. "It didn't have all the segments working," Zanoni admitted, but that did not stop him and Luce from showing it off at Optel's office Christmas party.[78] In recognition of his accomplishment, Luce was assigned to manage the LCD group. Zanoni, in turn, teamed up with Douglas Bosomworth and production coordinator Jack Heber to set up an LCD assembly line.[79] With these technical problems in good hands, Kiss launched a marketing drive. Because Optel possessed neither the brand recognition nor the manufacturing capacity to compete against existing watchmakers, he dispatched Kornstein and sales manager Gary Leffer to contact established firms eager to enter the digital watch market.[80] Bulova was an obvious starting point, but after supporting the development of the dynamic scattering watch, it instead placed its bets on the "Accuquartz" analog model.[81]

Thankfully, the Swiss were more open to Optel's overtures. In early 1971, Omega joined with the Americans to produce a "portable digital chronograph" called the Octoscope, which one Neuchâtel newspaper predicted would pave the way for "the 100% electronic watch."[82] Later that year, Optel formed an alliance with the Société des Garde-Temps (SGT), a conglomerate that had recently purchased the Waltham Watch Company, to produce watch movements and displays that would be inserted into Swiss cases. The results of this

Figure 5.4. "The Model T of digital watches." Optel's first commercially available LCD timepiece, built in partnership with the Waltham Watch Company, captivated audiences at the 1972 Basel Fair. (Courtesy of Zoltan Kiss.)

collaboration would then be marketed under the Waltham name in the United States.[83]

The Walchron, which Kiss declared "the Model T of digital watches," premiered to rave reviews at the 1972 Basel Fair, an annual showcase for the international watch and jewelry industries (fig. 5.4).[84] "That's the first time I experienced what it was like to be treated like a god," Kiss reminisced. "At the Basel Fair, there are all the watchmakers, and they found out that I was the guy who did this for Waltham, so there wasn't a single watchmaker that didn't besiege me."[85] Headlines in the Swiss press that spring referred to the Walchron as "a master stroke" and marveled at the idea of "a watch without gears."[86] They also applauded Optel, "a small company of 60 people. . . . A small business certainly, but one where 50% of the staff has a university education and the majority are from RCA, where they had a half-open window on watchmaking!"[87] On the basis of their joint venture with SGT, Optel soon found other European companies eager to team up with what one Swiss journalist dubbed "the young and restless business from New Jersey."[88]

With the firm reporting more than $10 million in orders, the time seemed ripe for expansion.[89] In June 1972, the company went public. Their initial stock offering earned them an additional $5 million, enough to cover the costs of moving out of its warehouse headquarters to a larger, more modern facility down Route 1.[90] Optel's relocation was the latest proof of how much the company had evolved over the past three years. As Kiss and his colleagues looked toward the second half of 1972, they could not imagine that greater changes were on the horizon or that the dynamic scattering devices that made Optel's fortune would soon face competition from a new form of LCD.

COMPETITION WITH A TWIST

In March 1972, as rumors swirled in the Swiss press about the dynamic scattering watches that Optel planned to unveil in Basel, Joseph Castellano published a review article surveying the present state of liquid crystal research.[91] Castellano had become leader of the DSRC's LCD group after George Heilmeier received his White House Fellowship. Zanoni and Goldmacher's move to join Kiss at Optel further diminished the DSRC team, but Castellano and his colleagues soldiered on, exploring a variety of electro-optic phenomena. He enumerated these effects in his 1972 article, beginning with dynamic scattering before moving on to guest-host color switching and the "optical storage mode" observed in nematic-cholesteric mixtures.[92]

Beyond its scientific content, Castellano's review illustrated how much interest in the topic had grown since Kent State hosted the first international liquid crystal conference in 1965. It cited projects at laboratories across the United States as well as in Britain, France, Germany, the Soviet Union, and Japan. Academic scientists and industrial researchers at places such as Bell Labs, Texas Instruments, Merck, and Hitachi were studying mesomorphic compounds and their applications.[93] There were now too many papers being published on liquid crystals to reference in a single article, which may explain the absence of one piece published the previous year by a former member of the DSRC technical staff named Wolfgang Helfrich.[94]

The omission was somewhat curious since Helfrich had served as part of RCA's liquid crystal team and, in fact, had shared an office with Castellano for several years.[95] Castellano specialized in organic synthesis while Helfrich, a physicist trained at the Technical University of Munich, modeled the behavior of liquid crystals under electric and magnetic fields in connection with the group's dynamic scattering investigations.[96] Because he was unfamiliar with liquid crystals before his 1967 arrival in Princeton, he spent much of his time in the DSRC library immersing himself in the relevant literature.[97]

It was there that Helfrich encountered the writings of a French crystallographer named Charles Mauguin. In a 1911 paper, Mauguin detailed a series of experiments confirming that one could align the rodlike molecules of a nematic compound by placing them on a glass plate that had been rubbed in a single direction with a piece of paper. Furthermore, if one formed a sandwich using two plates prepared in this fashion and rotated one of them ninety degrees, the molecules near each surface retained their original orientation while those in the middle contorted into a helical structure. When polarized light entered a cell prepared in this fashion, the helix behaved like a spiral staircase, rotating the light's plane of polarization on its way through the sample.[98]

Mauguin's findings had been publicly available for almost sixty years before Helfrich encountered them, but he was among the first to relate the optical properties of these "twisted nematic" systems to more recent studies of liquid crystals' electrical properties. After giving the matter some thought, he conceived of a new type of LCD that combined the apparatus used in Mauguin and Heilmeier's experiments (fig. 5.5, *left*). The RCA group had already shown that applying a voltage across certain liquid crystals caused the long axes of their molecules to realign parallel to the field and perpendicular to the

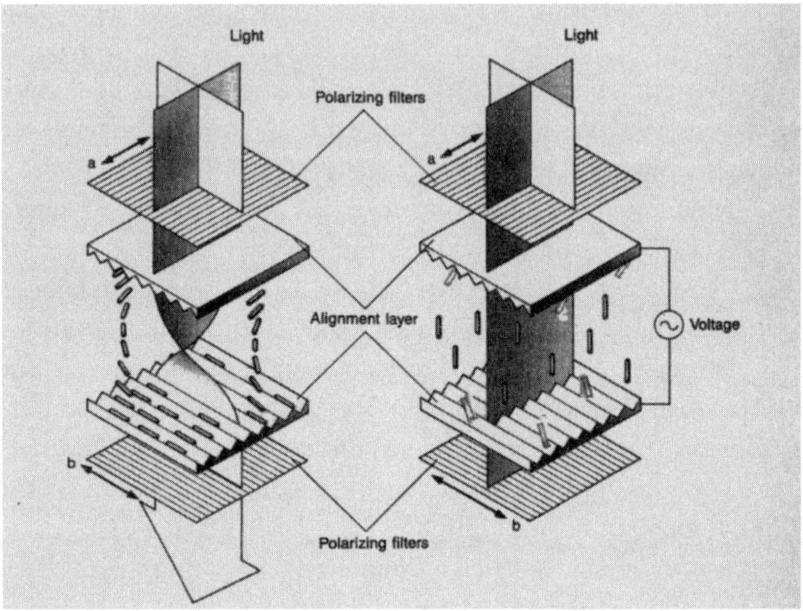

Figure 5.5. Wolfgang Helfrich proposed a new form of LCD, which involved creating a twisted nematic helix inside a sandwich cell and placing that setup between a pair of crossed polarizing filters. Each filter allowed only light waves moving in a particular direction, indicated by the arrows *a* and *b*, to pass through it. Before power was applied *(left)*, light traveling through the first polarizer followed the path of the liquid crystal helix, which enabled it to pass through the second filter. Applying a voltage to the cell caused the helix to unravel *(right)*. Light therefore moved across the cell with out changing direction and was blocked by the lower polarizer, making the display appear black. (Hirohisa Kawamoto, "The History of Liquid Crystal Displays," *Proceedings of the IEEE* 90, no. 4 [Apr. 2002]: 473. Reprinted with permission from IEEE. Scan courtesy of Linda Hall Library of Science, Engineering & Technology.)

substrates on which they were deposited.[99] These materials did not exhibit dynamic scattering, which required molecules whose long axes aligned perpendicular to the field.

Helfrich now envisioned using these nondynamic scattering materials to form a twisted nematic structure inside a standard RCA sandwich cell, which would be flanked by a pair of crossed polarizing filters. Normally, any light passing through the first polarizer would be blocked by the second, but the presence of a liquid crystal "staircase" between the two would rotate the light, allowing it to pass through the cell unobstructed until a voltage was applied across the sample (fig. 5.5, *right*). In that case, the field-induced molecular realignment would temporarily demolish the staircase, blocking all light transmission until the field was removed and the helix re-formed.

At least on paper, Helfrich had formulated an entirely new means of modulating light using liquid crystals, one that he recognized might have display applications. He later recalled presenting his idea to Heilmeier twice: once in the summer of 1969 and once in 1970. According to him, Heilmeier dismissed the proposal on both occasions. "And he gave a reason for that," Helfrich remembered. "It was the two polarizers. They absorb too much light, and therefore this is not an interesting effect. . . . I told him the whole concept, and he rejected it."[100]

Heilmeier's laboratory notebooks contain no record of these conversations, and when asked, he had no memory of speaking with Helfrich about the twisted nematic concept.[101] Whatever records Helfrich may have kept on the subject do not appear to have been deposited in the DSRC's archives, but there are several pieces of circumstantial evidence to support the claim that he conceived of the twisted nematic LCD at RCA. Technical reports indicate Helfrich's interest in the behavior of cholesteric liquid crystals, which are characterized by their helical structure.[102] Moreover, the operation of Helfrich's proposed twisted nematic cells somewhat resembled an electro-optic effect that other members of the DSRC group were studying in the spring of 1969, which used strong electric fields to unravel a cholesteric helix, causing it to shift from iridescent to transparent.[103] Finally, there are the recollections of other DSRC personnel, including Castellano, who witnessed Helfrich demonstrating "a cell that went from clear to dark in transmitted light under crossed polarizers when an electric field was applied."[104] Castellano points out that there is no way to know how that cell was constructed, but his experience confirms Helfrich's interest in using electric fields to modulate the passage of polarized light during his tenure at RCA.

Assuming Helfrich did approach Heilmeier with the idea for a twisted nematic LCD, how should we interpret the decision not to pursue it? Heilmeier's disdain for display systems that relied on polarized light was well established. The brightness reduction associated with polarizing filters had spurred his search for additional electro-optic effects following the observation of electronic color switching. Once dynamic scattering became the center of the DSRC group's attention, there were few incentives to return to displays based on polarized light, at least not without external funding, like NASA's contract to create color displays based on the guest-host effect.[105] From a broader perspective, by 1969, when Helfrich supposedly broached the subject of twisted nematic displays with Heilmeier, Princeton and Raritan's limited development budgets were stretched to the limit working on dynamic scattering devices. While it is possible that some of Heilmeier's reluctance to support Helfrich's proposal stemmed from a personal attachment to his dynamic scattering displays, it was also justifiable on technical and financial grounds.[106]

Helfrich had few responses to these objections. He was poorly positioned to advocate for a major reorientation of the LCD group's strategy, and the performance characteristics of his twisted nematic displays were largely theoretical. "The basics of the optical behavior were known," he noted later. "The basics of the electrical behavior were known, but the combination and in particular the light transmission of a distorted twisted nematic—distorted by the electric field—this was completely unknown and really difficult to estimate."[107] The answers to these questions would only be found in the fall of 1970, when Helfrich left RCA for a position at Hoffman–La Roche. The Swiss pharmaceutical firm recruited him to explore the mesomorphic properties of biological molecules and the possible use of LCDs in medical equipment. He joined their liquid crystal group in Basel—where Optel would later present its dynamic scattering watch—and partnered with experimental physicist Martin Schadt. After a few weeks of concerted tinkering, the two men successfully assembled a twisted nematic prototype based on Helfrich's idea in November 1970.[108] They submitted a Swiss patent application the following month and published their findings in a February 1971 article in *Applied Physics Letters*.[109]

Twisted nematic displays were more complicated to construct than their dynamic scattering predecessors. They relied on the installation of a pair of previously unnecessary polarizers, and the glass substrates required careful treatment to ensure a proper helical alignment. All the same, these devices offered several advantages. The distortion of the liquid crystal staircase in twisted nematic cells took place at lower voltages than dynamic scattering and did not rely on the flow of ions across a cell. (This absence of ionic current explains why some referred to the newer technology as simply a "field effect" display.) In addition, although the use of polarizers reduced their brightness, twisted nematic displays possessed higher contrast than dynamic scattering readouts.[110]

In years to come, these properties, enhanced with improved materials and fabrication techniques, enabled twisted nematic LCDs to dominate the electronic display industry. The now commonplace gray and black numerals seen on countless clocks and calculators would expand into the intricate arrays of light shutters found in our televisions, laptops, and cell phone screens. None of this was apparent when Helfrich and Schadt patented their invention. The enthusiasm within some sectors of the Swiss watch industry for digital timepieces was insufficient to persuade Hoffman–La Roche's management to support further LCD research. The firm ceased all liquid crystal investigations until inquiries from Japanese firms such as Seiko, Sharp, and Hitachi led to the LCD research group's reinstatement in 1973.[111]

Hoffman–La Roche's renewed interest in LCDs also drove them into an extended legal battle against James Fergason, the physicist who created the first practical temperature sensors using cholesteric liquid crystals. Fergason had independently developed his own twisted nematic displays and filed for a US patent two months after Helfrich and Schadt's Swiss application.[112] After several years of litigation, Fergason's Ohio-based start-up firm, International Liquid Crystal Company (ILIXCO) settled the dispute out of court. Fergason yielded control over his patents to Hoffman–La Roche in exchange for a share of the royalties. The Swiss firm then proceeded to license twisted nematic display technology to all interested parties.[113]

RCA was not initially among their number. Since Helfrich left no research records at the DSRC, the transatlantic wrangling over the twisted nematic patent did not affect Castellano or the Princeton liquid crystal group. Instead they continued applying for external funding and reaching out to RCA operating divisions with ideas for new dynamic scattering applications.[114] Meanwhile personnel at the Solid State Division (SSD) in Somerville put the finishing touches on an assembly line for dynamic scattering numeric indicators.[115] In February 1972, an SSD spokesman described plans to produce six types of dynamic scattering readouts for "minicalculators" and digital timepieces by the end of the year.[116]

Only that summer did DSRC scientists initiate a study comparing the performance of dynamic scattering and "field effect' displays under different viewing conditions. Along with technician Ronald Friel and electrical engineer Lawrence Goodman, Castellano surveyed more than two hundred DSRC employees, concluding that "a significantly large majority . . . preferred the cell with a specular reflecting back electrode using the dynamic scattering effect."[117] These findings validated the liquid crystal group's belief that consumers cared most about a display's brightness and their decision to avoid LCDs that relied on polarizing filters. The survey also indicated, however, that user preferences varied depending on ambient illumination. The white numbers in dynamic scattering displays were bright and legible under normal room lighting, but glare from their reflective backplates made them difficult to read outdoors. Conversely, the black digits of the twisted nematic cell were not as bright indoors but benefitted from increased contrast when viewed in direct sunlight.[118]

These findings contributed to the DSRC group's decision to place "major emphasis on materials for field-effect displays" for the remainder of 1972.[119] Nevertheless, Castellano and his colleagues continued research into dynamic scattering as competitors started abandoning the older displays.[120] To these firms, the decreased brightness and increased complexity of twisted nematic displays were small prices to pay for the high contrast and low power that made them perfect for portable electronics. By the end of 1973, Seiko had released a commercially successful twisted nematic wristwatch, and Optel started phasing out dynamic scattering models in favor of the new technology.[121]

As it became clear that dynamic scattering was falling out of favor with the industry, the DSRC and SSD liquid crystal groups began "an intensive program . . . to bring the twisted nematic field-effect display to commercial production."[122] Notwithstanding the economic slowdown following the Arab oil embargo, in February 1974, the *Wall Street Journal* reported that RCA was constructing a new LCD factory in Franklin Township, New Jersey.[123] The sixty thousand square-foot plant, dedicated to the production of twisted nematic clock and calculator displays, would employ one hundred people once it opened in October.[124] It was a move reminiscent of the establishment of the Raritan liquid crystal operation five years earlier, though at times it seemed RCA's management had learned little from that facility's closure.

To be sure, personnel at the Franklin Township facility resolved many technical issues associated with the fabrication of twisted nematic displays. Physical chemist Alan Sussman, who was hired by SSD after being let go from the DSRC in 1971, designed a machine that assured the proper alignment of liquid crystal molecules by depositing thin inorganic films on glass substrates instead of physically rubbing them.[125] That way "there was less handling. You didn't want to leave fingerprints. You didn't leave, you know, grit and stuff."[126] Sussman also eliminated a patchy, pox-like appearance that occasionally manifested in twisted nematic displays by adding a small amount of cholesteric material to induce uniform molecular orientation throughout the cell.[127] This technique remains standard practice among LCD manufacturers today.[128]

Much like at Raritan, these production hurdles were easier to resolve than those imposed by management. The newly appointed head of LCD development, receiving tube engineer Patrick Farina, chose to rely on veterans from RCA's previous assembly line to oversee daily operations.[129] Leading the way was "Lucky" Lawrence, who once more proceeded to alienate his subordinates among the technical staff.[130] This tactical blunder might have been tolerable had RCA's management not persisted in adhering to a contract-based funding strategy. DSRC engineer Lawrence Goodman—a frequent visitor to the new SSD facility—summarized the company's position: "If the Solid State Division could make a business out of watch displays and calculator displays, fine. But the main interest, corporate-wide was for television."[131] As before, the liquid crystal operation could not count on executive patronage.

If there was a silver lining to this gloomy financial situation, it was that the LCD had become more established, meaning that firms that had hesitated to approach RCA in 1969 were now more likely to invest in the technology. Sure enough, in 1975, the company secured a $4 million purchase order for twisted nematic readouts from Timex, America's highest-volume watch manufacturer, which saw the arrangement as a way of dipping its toes into the world of digital displays as it worked to establish its own liquid crystal and integrated circuit factories.[132] The Franklin Township plant, like Raritan before it, threw itself into fulfilling its contractual obligations, precluding discussion of any other LCD applications. So it was that RCA, the first-ever manufacturer of liquid crystal displays, was reduced to assembling numeric indicators derived from a technology it had previously shunned and selling them under someone else's name.

OPTEL'S SEARCH FOR STABILITY

As RCA came to terms with twisted nematic technology, its most prominent spin-off was undergoing rapid expansion. Between 1972 and 1975, Optel's workforce ballooned from thirty-three to 270 employees, and its revenues jumped from $170,000 to nearly $12 million.[133] This growth was fueled, at least in its early stages, by the vigorous pursuit of external contracts. Unlike RCA's leaders, who viewed liquid crystals as a diversion, Zoltan Kiss recognized that Optel's survival depended on his ability to drum up LCD-related business.

Working with his marketing staff, by the end of 1972 Kiss secured orders to produce over 250,000 watch movements for various Swiss-owned firms, including a call for 100,000 watches to be delivered over the course of three years to the Waltham Watch Company.[134]

Few disputed Kiss's aptitude as salesman, but as the Waltham contract made painfully clear, there was a vast difference between obtaining contracts and fulfilling them. Soon after announcing the deal in April 1972, Optel expanded its assembly lines with a goal of producing "between 20,000 and 100,000 movements by yearend."[135] The actual quantity delivered to Waltham was closer to 2,000 watches because of difficulties procuring CMOS chips. In a move that Kiss later characterized as naive, Optel had outsourced all of its integrated circuit needs to a single vendor, Solid State Scientific, which promised to produce the two chips required for each watch at a cost of $5.50 per pair. Shortly thereafter, Solid State determined that the circuits would be more difficult to produce than originally anticipated and raised the price to $10 a pair. Kiss protested the move, insisting that Solid State had an obligation to abide by its original agreement, but the absence of an alternative vendor and looming deadlines forced him to accept the chips at the higher rate.[136]

Unless Optel could find a cheaper source of circuits, it would barely break even in its dealings with Waltham. "To alleviate Optel's dependence on outside suppliers with respect to this key component," Kiss wrote in April 1973, "we have hired during the past quarter the technical personnel necessary to design and generate our own masks for proprietary integrated circuit chips."[137] Optel later announced that New York–based Solitron Devices would step in to fill the void left by Solid State Scientific, though Kiss and his colleagues also reached out to other firms, including RCA, to ensure redundancy in their supply chain. These steps proved sufficient to appease Waltham, which had deliberated suing Optel for failing to reach its production quotas.[138]

Resolving the semiconductor supply question did not settle other technical issues with Optel's watches. In November 1973, Optel executive vice president David Barnett confessed to shareholders that one-quarter of the company's timepieces were being returned due to problems with the display, the chips, or "a range of other reasons."[139] Barnett, formerly a director of engineering at Sperry Rand, clarified that this return rate was nothing unusual. "These problems are no more and no less than have been experienced in other instances of a new technology being applied to a new and demanding market," he asserted, before admitting that "to achieve the kind of goal we have set for ourselves, improvements in product reliability will be required."[140]

As 1973 drew to a close, Optel took several important steps to address these problems. It expanded its integrated circuit design capacity and improved display construction protocols. The transition to twisted nematics was also underway. Kiss framed that shift as an aesthetic choice, but the newer displays' lower power requirements also allowed Optel to reuse chips previously deemed unsuitable for dynamic scattering equipment.[141] None of these developments stabilized the company's financial performance. Optel's revenues were increasing, but so were its expenses. The firm recorded its first profitable quarter in September 1973, but its net losses for the year totaled nearly $1.7 million.[142]

Optel's official statements blamed these losses on technical difficulties and "delays in receiving watch cases and other parts," but some inside the company felt the guilt should be placed squarely on the shoulders of Zoltan Kiss.[143] "He was a very, very brilliant man, but he didn't know how to run a company," Tony Luce explained. "He just didn't know how to make money. That was his problem."[144] Few doubted Kiss's scientific credentials or his sales acumen, but the handling of the Waltham contract betrayed his lack of administrative experience. Of equal concern was his tendency to overstate Optel's manufacturing capacity. Edward Kornstein lashed out at Kiss for including "all kinds of numbers on production rates and so on and so forth that were absolute nonsense" in the company's public offering statement.[145] Kiss's insistence on retaining these exaggerated figures in investor presentations finally prompted Kornstein, one of his long-time critics, to leave Optel.

Until the end of 1974, Kiss was able to stifle these attacks. He

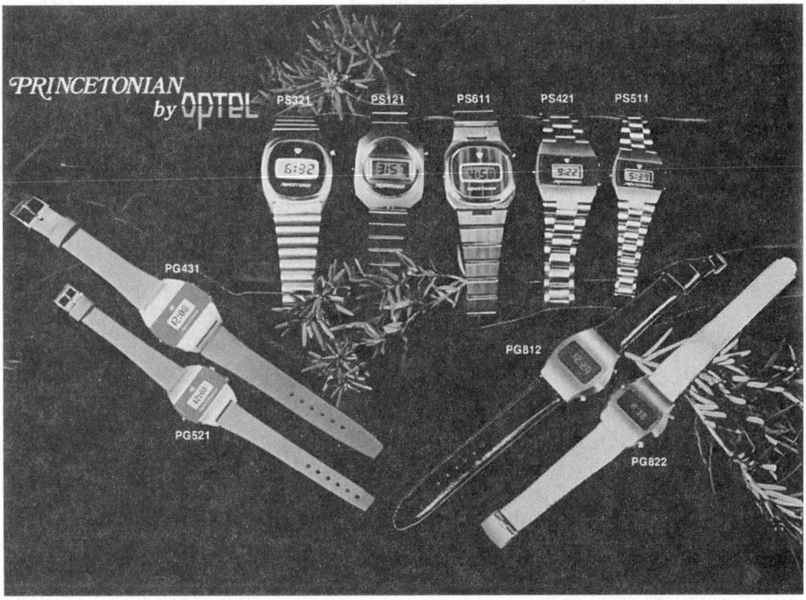

Figure 5.6. Advertisement for Optel's Princetonian watch line featuring gray-and-black twisted nematic displays. The two watches in the lower right corner (PG812 and PG822) had red LED readouts. (Courtesy of Louis Zanoni.)

bolstered his position by arranging a deal with American Express to distribute Optel's Quartz Segtronic watch via mail order.[146] Cardholders could purchase this "computer timepiece," featuring a twisted nematic display and a stainless steel case, for $119 or upgrade to a "luxurious electroplated 18kt gold" version for $149.[147] Buoyed by the success of the American Express negotiations, Kiss moved forward with plans to supplement Optel's contract income by producing its own watch lines. Once again, Luce remembered, "Zoltan started to see the dollar signs, he's saying, wait a minute, this is not right, we have to go out there and start selling watches rather than waiting for the Omegas or Bulovas or Timexes to come and say, make us the watch. . . . So he just went ahead and did that."[148] Soon Optel was marketing timepieces under its own name and planning its new Cadrille and Princetonian watch collections (fig. 5.6).

The transformation from contractor to independent manufacturer, effectively the reverse of RCA's handling of the LCD, reflected Kiss's confidence in Optel's prospects but alienated several of the firm's clients. Kornstein, who had moved on to become a watch industry consultant, observed that companies like Waltham "needed a source of watch modules and Optel was not interested in providing them with modules anymore. Optel was trying to establish their own watch name, which they did."[149] The income loss connected with this move precipitated a pair of major decisions in spring 1974. First, at the end of March, Optel obtained a $1 million loan from the Chemical Bank of New York backed by a consortium of Swiss watch firms.[150] This news provoked speculation that a change of management was forthcoming. As predicted, the following week Zoltan Kiss stepped down as president, replaced by none other than his former DSRC supervisor, Henry Lewis.[151]

Optel's directors believed that Lewis, who had left RCA in 1970 for an executive position at Itek Corporation, might be able to rein in Kiss, who stayed on as Optel's chairman. These hopes were misplaced. Within five months Lewis resigned his position, and Kiss reassumed Optel's presidency.[152] On the surface little changed at Optel during its founder's brief fall from grace. The company's losses continued to outpace its swiftly growing revenues, leading it to solicit $1.5 million in additional financing from Delta Transnational, a subsidiary of Mitsubishi.[153] Publicity literature framed this infusion of capital as a means of "assuring the stability of Optel during its transition from a research and development organization to a consistently profitable full-capability watch manufacturer," but following Lewis's resignation, things looked anything but stable to the members of the firm's board.[154] Faced with an ever more crowded market for digital watches, Optel's leaders pressed for an investment in LEDs, which were widely seen as more reliable than liquid crystals.[155]

The appeal of LEDs eluded Zanoni, who had, a decade earlier, built the first dynamic scattering displays. As he put it, "we never realized that people would buy it and pay money to press a button to tell time when we could just look at our watches to tell time."[156] Just the same, Optel soon revived its partnership with American Express to promote the Ultimatic, an LED watch touted as one of the first quartz timepieces with separate models for men and women.[157] The new emphasis on LEDs did not prevent Optel's technical staff from improving its LCD product lines. The firm was an early adopter of "zebra connectors," flexible strips consisting of alternating conductive and insulating layers that replaced soldered linkages between a watch's display and circuit board.[158] This advancement facilitated the introduction of more complex timepieces, including the Optel I, a calculator wristwatch exhibited at the 1975 Basel Fair.[159]

These technical advances did little to ward off Optel's financial woes. Even after receiving another $1 million loan from Chemical Bank and canceling $2.5 million of long-term debt by transferring 40 percent of its common stock to Delta, the firm registered a $4 million net loss in 1975.[160] The board of directors turned its eyes to Kiss and decided that another change in leadership was in order. In February 1976, they announced that Gerald Heller, the leader of an instrumentation firm called ILC Data Device, would take over as president and CEO of Optel.

When questioned about his plans for the company, Heller told reporters that Optel would "explore display technologies to their maximum," with a particular focus on LCDs and new electrochromic materials, which changed color under an electric field.[161] The firm would also concentrate on producing high-end timepieces, leaving larger companies like Texas Instruments and Fairchild Semiconductor to fight over the low-cost watch market. Surprisingly, Kiss remained as Optel's chairman "with prime responsibility for research and development."[162] But unlike during Lewis's presidency, when he retained the title of chief executive, he now had little sway over long-term strategy. For the first time since Optel's founding, Zoltan Kiss was no longer master of the company's fate.

THE GENERAL'S DREAM REVISITED

The sense of powerlessness that Kiss faced at Optel would have been recognizable to the liquid crystal researchers he left behind at the

DSRC. Industry sources estimated that RCA invested $9 million in LCD manufacturing between 1970 and 1976, but little of this money trickled back to Princeton.[163] Believing that further research was not necessary to improve the numeric readouts that had become RCA's main LCD product line, the DSRC's management briefly terminated the liquid crystal project in the summer of 1972. The cancellation lasted only two days thanks to personnel in the company's patent division, who pointed out that additional technical support could assist in its campaign to license LCD technology to other firms. The incident nonetheless cast a pall over the shrinking research group, prompting its leader, Joseph Castellano, to leave RCA and join a new LCD start-up, Princeton Materials Science, in the spring of 1973.[164]

Ironically, the timing of Castellano's departure coincided with the resurrection of General Sarnoff's old idea of a television that could hang on the wall. The first steps in this direction were taken at a series of meetings inaugurated by William Webster. The turmoil surrounding RCA's annus horribilis had deeply affected the DSRC head, who reportedly aged visibly during the final months of 1971, and he resolved to protect the Princeton labs by forming stronger alliances with the rest of the corporation.[165] To that end, in early 1972 he organized a series of planning sessions where corporate staff and representatives from the operating divisions evaluated all ongoing research in Princeton "with respect to such factors as risk, potential payoff, and timing."[166] Based on these conversations, the DSRC launched several consumer electronics projects, including a "new program with the goal of a flat panel display for TV receivers."[167]

Like his father two decades earlier, Robert Sarnoff did not wait for the completion of a prototype before promoting RCA's latest flat-panel television project. At a 1973 shareholders meeting in Dallas, he boasted about rising earnings and the ongoing development of "luminescent flat-screen television."[168] When asked about the future of consumer electronics during an in-house interview later that summer, Sarnoff referred to the notion of "a flat, luminescent display screen that could go on the wall" as "a high priority with us."[169] RCA researchers, he continued, "have been working in this area for nearly

30 years," presumably a reference to projection systems used in the 1940s.[170]

In that same interview, Sarnoff referred to several "new and promising technologies" that could make flat-panel television possible, but perhaps taking a lead from his father's Magnalux speech, he left the details to his readers' imaginations. This new flat-panel project may have had support from the operating divisions and 30 Rockefeller Plaza, but as always the technological basis would depend on assessments from the DSRC's technical staff. In July 1972, Webster had selected John van Raalte, someone well acquainted with previous RCA attempts to replace the CRT, to assemble a "Flat TV Display Task Force" consisting of "10–15 RCA Laboratories scientists with diverse technical backgrounds." This group would ascertain "the most promising approach towards the development of a large-area flat, consumer, color TV receiver."[171] Such a device had to match the image quality and power requirements of existing televisions while expanding the screen's diagonal measurement to 50 inches.[172]

Van Raalte's task force spent the second half of 1972 evaluating candidate technologies for use in this hypothetical display. In spite of promising RCA material science investigations that resulted in the first blue light-emitting diode, which in combination with existing red and green LEDs might facilitate a full-color display, the high cost and low brightness of these devices disqualified them from further consideration.[173] Plasma displays could emit light across the entirety of the visible spectrum by exciting phosphors with ultraviolet light, but the process was inefficient and relied on compounds that tended to decompose. Electroluminescence, a focus of DSRC flat-panel research since the 1950s, was similarly eliminated for failing to match the efficiency and reliability of RCA's television phosphors.[174]

Alongside these emissive display technologies, which produced their own light, the DSRC task force also reviewed so-called passive displays, which modulated externally generated light.[175] Liquid crystals fell into the latter category, and Van Raalte assigned Lawrence Goodman, who had studied matrix-addressed LCDs as a member of Castellano's group, the responsibility of summarizing the strengths and weaknesses of a liquid crystal television.[176] Goodman laid out the case for liquid crystals, highlighting their low power, fast switching speeds, and compatibility with large-area fabrication. He also echoed Bernard Lechner's conclusion that each of the 250,000-plus picture elements in the proposed display would require its own diode or transistor to ensure proper switching.[177] Goodman had previously researched the possibility of using thin-film devices to perform this function, but the difficulty of scaling up their production, as well as restrictions on illumination, color, and viewing angle, did not bode well for the LCD.[178]

After several months of deliberation, in December 1972 the Flat TV Display Task Force submitted a report to Webster and the rest of the DSRC's management. They had weighed all of the alternatives to the cathode-ray tube and found them wanting. None of the technologies considered, including liquid crystals, could match the luminous efficiency, peak brightness, or color gamut of commercially available televisions. "It was basically for these reasons," Van Raalte wrote in a memorandum the following year, "that the Task Force recommended that RCA initiate a major research effort to develop a flat 30" × 40" cathodoluminescent TV display."[179] Rather than replace the CRT with an unproven technology, Princeton researchers would redesign one that had withstood the test of time, adding a new internal support structure and replacing the electron gun with a "large-area cathode" that would shower electrons across the entire surface of the display.[180]

Van Raalte emphasized that the task force's report was not a call to terminate research into alternative display technologies. "For the small, monochrome display application," he noted, "both plasmas and DC-EL [direct current-electroluminescence] have already demonstrated near-acceptable performance."[181] For the rest of the 1970s, however, the bulk of flat-panel research support at the DSRC went toward construction of a modified CRT, which replaced the large-area cathode concept with a series of guided electron beams. The project resulted in a functional prototype before its cancellation in the late 1970s to secure additional manpower for the VideoDisc initiative.[182]

If Van Raalte's inquiries and the resulting CRT-based projects confirmed the persistence of David Sarnoff's dream of a wall-mounted television, they also laid to rest any further RCA discussion of an LCD television. Liquid crystal investigations continued at the DSRC after Castellano's move to Princeton Materials Science. Under the leadership of Daniel Ross, the chemist who years before had supplied Heilmeier with dyes for his first guest-host color switching experiments, the LCD group stayed involved in display development but only insofar as such work supported SSD's "push to success in this intriguing but difficult technology."[183] Over the next few years, the group devised improved materials and assembly techniques for Somerville's new line of twisted nematic watch displays.[184] Television applications were no longer in the picture.

Without flat-panel TV as an end point, liquid crystals reverted back to relative obscurity, subsumed within RCA's broader research agenda. It is noteworthy that the most prominent application of liquid crystals to come out of the DSRC between 1970 and 1976 had absolutely nothing to do with displays. Instead, RCA shareholders and readers of the *New York Times* learned about physicist Donald Channin's creative use of nematic compounds to test the functionality of integrated circuits (fig. 5.7).[185] The DSRC's leaders had once hailed liquid crystals as RCA's best hope of creating a wall-mounted television. Now they were content to sustain a marginal LCD research effort to support SSD's factory in Franklin Township—nothing more. As it turned out, even this limited goal would prove overly ambitious.

A SOLUTION LOOKING FOR A PROBLEM

Rumors that RCA was negotiating the sale of its liquid crystal operation reached the press in February 1976, three months after Robert Sarnoff stepped down as the corporation's chief executive.[186] Simmering frustration with Sarnoff's leadership style and the steady erosion of profits since 1973 had culminated in what the press termed a "palace revolt" at the highest levels of RCA's management.[187] Upon taking office, veteran administrator Anthony Conrad—the first

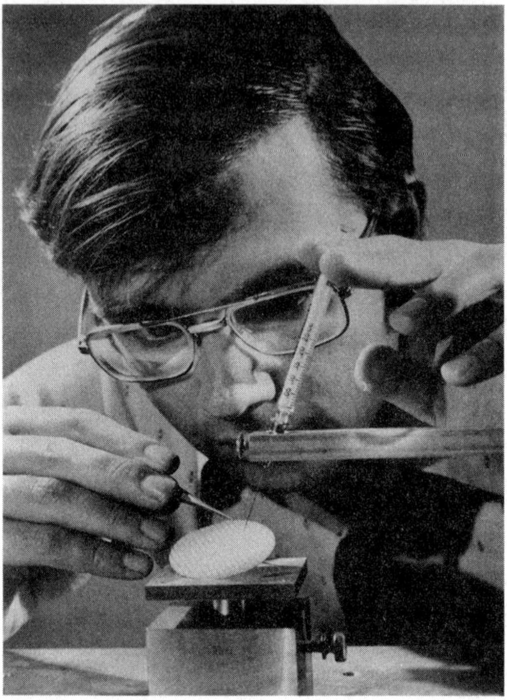

Figure 5.7. By the mid-1970s, RCA was no longer promoting the use of liquid crystals in displays. The most prominent application they publicized was a technique developed by DSRC physicist Donald Channin (shown here) to observe the operation of integrated circuits with a layer of nematic materials. (David Sarnoff Library Collection, courtesy of Hagley Museum and Library.)

non-Sarnoff to run RCA in a generation—launched another round of cost-cutting measures, beginning with the company's solid-state and picture tube divisions, each of which had posted multimillion dollar losses in 1975.[188] The Franklin Township LCD factory was among the items that SSD placed on the chopping block. In April 1976, the division's vice president, Bernard Vonderschmitt, finalized a deal with Timex offering them the entire liquid crystal operation and access to RCA's sizable portfolio of LCD patents for $2.45 million.[189]

Any sense of schadenfreude that Zoltan Kiss and the RCA alumni at Optel may have felt toward their corporate parent was short lived. Two weeks after the conclusion of the Timex deal, Kiss resigned from

Optel to establish a new solar energy start-up called Chronar.[190] By then, Gerald Heller's efforts to salvage Optel's finances had exacted a heavy toll, with one-third of the company's 450-person workforce losing their jobs.[191] The resulting savings were not enough to keep Optel from reaching out to Wall Street for an emergency infusion of working capital.[192] Optel's position deteriorated further at the 1976 Basel Fair, where one of its Swiss partners threatened to physically block the opening of the firm's booth unless it repaid its outstanding debt.[193] Other European watchmakers followed suit, and soon Optel found itself on the hook for roughly 10 million francs.[194] Heller negotiated a reprieve until early June, at which point Chemical Bank, one of start-up's largest shareholders, joined the chorus of creditors demanding payment on its loans. The company had no choice but to file for bankruptcy.[195]

The consequences of the near simultaneous collapse of liquid crystal operations at RCA and Optel in the spring of 1976 reflected the contributions of LCD technology to each firm's respective bottom line. RCA, a sprawling conglomerate whose annual revenues were about to exceed $5 billion for the first time, absorbed the loss with minimal difficulty.[196] Whatever disappointment the news may have triggered among the DSRC's staff was tempered by an awareness of SSD's financial weakness and longtime lack of management support.[197] At Optel, the effects were far more dire. The declaration of bankruptcy did not stop the company from producing new watches, including its first solar-powered model, but it undermined several promising refinancing schemes.[198] To ensure a partial return on its shareholders' investments, in 1978 Optel merged with discount drugstore operator Levitt Industries.[199] Shedding its previous identity as a manufacturer, it sold off its LCD-related assets to Refac Electronics, a licensing firm that earned millions from its newly acquired intellectual property over the ensuing decades.[200]

In death as in life, Optel's behavior diverged wildly from RCA, but the two firms remained united in their inability to capitalize on early entry into the liquid crystal display industry. RCA could never dispel the conservative miasma that descended on the company following the 1971 computer collapse, while Optel's enthusiasm for LCDs led to rapid but untenable growth. Both companies struggled to delineate the market for their displays and the products most likely to attract consumers. Zoltan Kiss, alluding to these difficulties, later referred to liquid crystals as "a solution looking for a problem."[201] Today we might characterize the LCD as a disruptive technology.

The adoption of Christensen's terminology encourages us to consider who determines whether a technology is disruptive. As we have seen, Heilmeier and his colleagues shaped managerial expectations of the LCD from its inception through pilot production. Researchers at RCA and Optel adopted a similar role as their displays reached the marketplace in the early 1970s, basing their recommendations on their growing familiarity with the capabilities and limitations of liquid crystals—their material logic. To take one example, the switching properties of nematic compounds and the sheer amount of circuitry required to incorporate them into a matrix-addressed display were key factors in the Flat TV Display Task Force's consensus that RCA should not pursue the construction of a liquid crystal television. Similarly, Optel's proposal to develop an electronic wristwatch for Bulova relied on the assertions of experts such as Luce, Zanoni, and Goldmacher that a dynamic scattering readout could be developed that was compatible with existing integrated circuits, something that their colleagues at RCA dismissed in the short term.

More significantly, the juxtaposition of the RCA and Optel case studies reminds us of the fluidity of Christensen's categories and the context-dependent nature of technological disruption. A firm's research capacity, fiscal commitments, and institutional norms all shape how its technical staff will evaluate a new technology. By way of illustration, consider dynamic scattering, which fulfilled all of the characteristics of a disruptive technology upon its introduction. Dynamic scattering LCDs did little to enhance RCA's existing businesses but offered features, such as portability and low operating power, of potential value in other applications. The DSRC liquid crystal group recognized these advantages and spent years lobbying their colleagues to invest in products that benefitted from them.

Yet these same personnel bristled at the introduction of the twisted nematic display. Suddenly the roles were reversed, with the onetime disruptors now opposing disruption. Twisted nematic displays possessed lower operating voltages and improved contrast, but to Heilmeier and his colleagues they represented a step backward, whose reliance on polarized light implied a more complex construction process and dimmer images. This initial resistance prevented Helfrich from pursuing the development of the twisted nematic display at RCA and slowed the firm's subsequent adoption of what became a far more profitable technology than dynamic scattering.

At Optel, the acceptance of twisted nematic displays was less tumultuous. Making the switch required new equipment, procedures, and materials, but Optel scientists recognized that these new LCDs were well suited for the watch displays that had become the company's livelihood. With so much on the line for the young spin-off, even former members of Heilmeier's research team learned to stop worrying about polarized filters and love twisted nematics. In other words, what was treated as a disruptive technology at RCA became its opposite—a sustaining technology—at Optel. The smoothness of this transition was not guaranteed. That the firm nearly tore itself apart over the decision to prioritize liquid crystal displays instead of cathodochromics challenges Christensen's assertion that disruption comes easier to start-up firms.[202]

The permeability described here lends further credence to historian Jill Lepore's 2014 observation that "disruptive innovation can reliably be seen only after the fact."[203] More broadly, it calls into question the utility of "disruption" and "disruptive innovation" as analytical categories. No one at RCA or Optel, after all, used that vocabulary to describe their display research or any products that resulted from it. The business press referred to "competition," "chaos," "shakeouts," and even a "revolution" in the digital watch industry, but never "disruption" in Christensen's sense.[204]

Still, if we are careful to avoid anachronism, this framework provides a useful means of talking about how corporations compare technologies and allocate resources toward their commercialization. Strip away the more problematic, normative aspects of his theory and essentially Christensen is telling a story about managers making tough decisions, often with input from "non-executive participants"—such as scientists and engineers—based on their understanding of the market, the company's product lines, and their own career priorities.[205] The results of these deliberations, as Lepore rightly points out, are notoriously difficult to apply from one situation to another. That does not render those discussions less important to understanding the settings in which technologies like the LCD, which depend on new and unfamiliar materials, will prosper.[206]

Liquid crystal displays did not find a home at RCA, an outcome that George Heilmeier later attributed to managers who saw them "more as a threat than an opportunity."[207] The project might have succeeded, he argued, if the team behind the creation of the LCD had "been given the responsibility for developing the business opportunity as well as the technology."[208] In a way, Optel embodied this road not taken. And for a few years, it appeared Heilmeier was correct, as Zoltan Kiss created a space where LCD development could take place unencumbered by any preexisting corporate commitments. Optel's ultimate collapse, however, showed that technical skill alone could not guarantee an innovation's success. Only by striking a balance between scientific and managerial expertise could one nurture an LCD application from a prototype into the basis for a commercially sustainable enterprise. In the end, neither company would accomplish that task. As Optel joined the ranks of influential but unsuccessful high-tech start-ups, RCA's scientists could only watch as the inventions that ultimately allowed the dream of a flat-panel television to become a reality left Princeton for greener pastures, in the United States and beyond.

CONCLUSION: AN INVISIBLE MONUMENT

To drivers passing through Princeton on US Route 1, the birthplace of the LCD looks much the same as it did when David Sarnoff wandered its halls. The RCA insignia has long since vanished from the front lawn, replaced by a smaller blue sign confirming the laboratory's new identity as a branch of SRI International, a nonprofit scientific research organization. General Electric acquired the facility in 1986 as part of the $6.3 billion deal to purchase RCA following the commercial failure of the VideoDisc.[1] At the time, it was the largest merger in American history that did not involve an oil company, but GE executives saw no reason to maintain two corporate research centers and donated the DSRC to SRI in 1987.[2] Under new leadership, former members of RCA's technical staff remained key players in the consumer electronics industry, most notably through membership in the "Grand Alliance" responsible for the American high-definition television (HDTV) broadcasting standard. They also expanded investigations into areas of greater interest to industrial and government clients, including biomedical systems and computer vision technologies.[3]

The DSRC building weathered these changes in management, institutional affiliation, and research priorities with few updates to its external appearance. Of course, there were alterations, but observing them requires drivers to pull off the highway, turn down a tree-lined side street, and park near the marble and glass portico that has served as the main entrance since 1966. This entryway, still inscribed with the meatball logo of its former owner, contains some of the newest additions to the research center's façade: a series of bronze plaques commissioned by the history committee of the Institute of Electrical and Electronics Engineers (IEEE). These "milestones" commemorate RCA's development of compatible color television, the construction of the TIROS meteorological satellite, and the invention of the liquid crystal display.[4]

The LCD milestone, dated September 2006, is embedded on the right side of the doorway as you enter the building. Except for a diamond-shaped IEEE logo in the bottom-left corner, its contents are entirely textual. In addition to two lines indicating its provenance, the plaque includes a title ("Liquid Crystal Display, 1968") and a summary of the research being honored:

> Between 1964 and 1968, at the RCA David Sarnoff Research Center in Princeton, New Jersey, a team of engineers and scientists led by George H. Heilmeier with Louis A. Zanoni and Lucian A. Barton devised a method for electronic control of light reflected from liquid crystals and demonstrated the first liquid crystal display. Their work launched a global industry that now produces millions of LCDs annually for watches, calculators, flat-panel displays in televisions, computers, and instruments.[5]

There is an undeniable elegance to this milestone's summary, which compresses the complex origins of the LCD down to a pair of sentences. It presents a geographically and chronologically bounded narrative of technological innovation and winnows down the dozens of contributors to the creation of the liquid crystal display to a far more manageable trio. Yet as much as Heilmeier, Zanoni, and Barton deserve recognition for their achievements, the narrative preserved on the wall of their former laboratory leaves many questions unanswered. A thoughtful visitor might well ask what inspired these researchers to examine liquid crystals and how many other people were members of their team. She might also wonder what happened to the RCA liquid crystal project after 1968 and the extent of the company's involvement with the present-day LCD industry.

The answer to at least one of these questions is hidden in plain sight within SRI's Princeton facility, though its connection to liquid crystals is not obvious. Only someone familiar with RCA's history would recognize the second plaque hanging on the far side of the lobby within eyeshot of the IEEE milestone. Capped with a bronze relief image of David Sarnoff, this plaque celebrates the September day in 1951 when RCA renamed the research center in the General's honor. Sarnoff's speech that afternoon prompted the company's first serious attempts to develop a flat-panel television, triggering the cascade of events that ultimately led to the LCD.

These two plaques, dedicated fifty-five years apart, serve as imperfect bookends to the story of liquid crystal research at RCA. Even if our hypothetical visitor were able to connect the dots between Sarnoff's request for Magnalux and the assembly of the first dynamic scattering displays, the IEEE milestone provides scarcely any details about the company's post-1968 accomplishments. It is easy to chalk up this absence to the precipitous collapse of RCA's liquid crystal operation, but there is also an argument to be made that the DSRC group's most lasting contributions to consumer electronics occurred only after they took the LCD public.

RCA scientists and engineers had already proven their ability to alter how personnel within their company perceived various flat-panel display technologies. The revelation of their prototype LCDs performed an analogous function but on a grander scale. As news spread of liquid crystals' unique properties, other companies—first in the United States and then overseas—established LCD research programs of their own. Directly and indirectly, RCA and its technical staff transformed how these businesses approached the commercialization of liquid crystals. The absence of a physical monument honoring their role in the dissemination of the LCD does not diminish its significance.

• • •

Unlike RCA's managers, whose enthusiasm toward liquid crystals faded in the wake of the 1968 press conference, their peers at other US electronics firms viewed these new materials as an exciting opportunity. The relative obscurity of mesomorphic compounds meant that they had to rely on articles written by RCA scientists for preliminary information about their electro-optic properties. Sun Lu, the electrical engineer responsible for organizing Texas Instruments' LCD program, specifically recalled reading Heilmeier's paper on dynamic scattering before recruiting the first two chemists to his group.[6] Other companies, recognizing the importance of firsthand expertise fabricating LCDs, took the additional step of hiring members of the DSRC liquid crystal project. Several years before purchasing RCA's display factory in Franklin Township, Timex persuaded organic chemist Chan Soo Oh, who had joined Heilmeier's team shortly before the engineer's 1970 exit, to build up LCD production at their R & D center in Tarrytown, New York.[7] Oh's colleague, Joseph Castellano, would also find his services in demand once Sprague Electric, a leading capacitor manufacturer, became majority shareholder of Princeton Materials Science at the end of 1973. Castellano worked closely with Sprague executives to facilitate the start-up's transition from dynamic scattering to twisted nematic displays.[8]

Eventually, both Castellano and Oh traveled to California in response to rising entrepreneurial interest in liquid crystals. In 1974, Castellano received an invitation from Robert Noyce, the CEO of Intel, to serve as a liaison between the semiconductor manufacturer and Microma, its recently purchased LCD subsidiary. He deferred because of ongoing commitments at Princeton Materials Science, but the following year Sprague sold the firm's assets to Fairchild Semiconductor. The integrated circuit pioneer was already making LED watch displays and hired Castellano to install a parallel LCD production line.[9] Oh, meanwhile, accepted an offer from Beckman Instruments, which was seeking low-power readouts for its battery-operated pH meters.[10] Organic chemistry, previously consigned to the periphery of the electronics industry, was coming into its own, and the knowledge of liquid crystalline materials accumulated over years at the DSRC migrated westward.

For a little while, these West Coast liquid crystal operations flourished. Under Castellano's supervision, Fairchild increased the efficiency of its assembly lines until they were churning out thousands of displays each day.[11] At the same time, Oh put together an organic synthesis laboratory at Beckman and evaluated new display fabrication and sealing techniques. By the end of the 1970s, Beckman had become one of the world's leading producers of LCD readouts for scientific instruments, toys, and wristwatches.[12] But for Beckman and Fairchild, as well as the numerous Silicon Valley LCD firms that sprang up during the 1970s, such success proved short lived. The apparent synergy between liquid crystals and other high-tech industries could not surmount the proportionally high labor and material costs associated with LCD manufacturing. The fact that many of their displays were used in low-cost products aimed at the mass market, such as wristwatches, only exacerbated the problem.[13]

In response to these pressures, Beckman phased out its liquid crystal business, though Oh was allowed to remain head of the organic chemistry group in the company's Diagnostic Division.[14] Fairchild stayed in the market but shifted all of its LCD production to Hong Kong, where lower wages promised a decrease in costs. Castellano helped design the company's new factory before taking a job with an Exxon LCD subsidiary (Datascreen Corporation, later Kylex) and eventually becoming a full-time display industry consultant.[15]

Not even Intel, hardly a slouch when it came to implementing new technologies, could justify maintaining its investment in Microma. Gordon Moore, Intel's cofounder and future chairman, had anticipated digital timepieces in 1965, but now rivals such as Texas Instruments were showing that success in the watch business relied as much on savvy salesmanship as technical aptitude.[16] All the while, Intel's move into watchmaking was falling victim to Moore's namesake law. "We misread the way the business was going," he recalled. "When we got out, the semiconductor content, the chip, cost less than the little pins on the side of the case."[17] With at least sixty US companies engaged in digital watch production and prices plummeting, Intel cut its losses in September 1977.[18] It sold the Microma brand to Endura, a Swiss manufacturer, and its watch factory in Cupertino to Timex.[19]

As these California liquid crystal operations declined, Timex accelerated its adoption of the technology in a market previously split between LED and LCD readouts. In spite of their shorter battery life, LEDs had retained a slight lead among watchmakers during the first half of the 1970s due to their perceived reliability compared to early liquid crystal models. Timex initially avoided taking sides on the issue, preferring to focus its attention on mechanical timepieces until Texas Instruments forced its hand by introducing a $20 LED watch in 1976 and a $10 version a year later.[20] The moment had come for action, so Timex made its move. They doubled down on liquid crystal wristwatches, whose "always-on" readouts had started to win over consumers.[21]

More than market forces underlay Timex executives' decision to adopt the LCD. Their interactions with personnel at RCA's Franklin Township plant also shaped their understanding of liquid crystals' material logic and would continue to do so after they purchased that facility in 1976.[22] That deal, along with the Microma acquisition, sparked widespread doubt about the long-term prospects of the LED wristwatch. Fearing that a major shift toward liquid crystals was imminent, Texas Instruments and National Semiconductor began considering LCDs more seriously. "By the end of 1977," one early analysis of the American LCD industry noted, "nearly all LED inventories were either written off or sold below cost through dumping and give-away programs."[23] Timex invoked their old slogan in advertising campaigns that vowed their new LCD watches would "take a lickin' and keep on tickin'" even without a mechanical movement.[24] With the endorsement of the industry's undisputed leader, the LCD guaranteed its position as the prevailing electronic wristwatch technology in the United States.

American LCD manufacturers had little time to savor their victory. The rapidly falling prices of digital clocks forced most firms to drop out of the industry, leaving Fairchild, Timex, and Texas Instruments to jockey for supremacy.[25] Though their respective market shares increased, the razor-thin margins associated with the watch business made it difficult to boost profits. Despite a concerted effort to promote its high-end, luxury timepieces, Texas Instruments announced plans to phase out all LCD watch manufacturing in the spring of 1981.[26] As previously mentioned, Fairchild held out slightly longer by moving its liquid crystal facilities to Hong Kong, and Timex later adopted a similar tactic. They closed the former RCA and Microma plants in New Jersey and California and relocated watch fabrication to Hong Kong, Singapore, and Taiwan before consolidating production in the Philippines.[27]

The US flat-panel display industry never truly recovered from the loss of these liquid crystal operations. At the time, policy makers expressed little concern about their departure. Surprisingly, one of the few people who might have been able to reverse this trend was George Heilmeier. At the conclusion of his White House fellowship, Heilmeier pursued a career in the Defense Department, rising through the ranks to become director of the Defense Advanced Research Projects Agency (DARPA) in 1975. From this position, Heilmeier authorized the funding of stealth aircraft and artificial intelligence projects but consistently turned down opportunities to support flat-panel display research.[28] He defended his position on pragmatic grounds, encapsulated in a keynote lecture delivered at a 1972 IEEE conference:

> How many realistic scenarios are there in which we win because we have a flat-panel, matrix-addressed display in the cockpit while the other guy has a CRT? And how many would we buy if they did exist?[29]

Upon leaving DARPA in 1977, Heilmeier would have additional opportunities to sponsor liquid crystal display projects as the vice president and chief technology officer of Texas Instruments. Time and again, however, he passed over the LCD, even going so far as to encourage the use of LED readouts in the company's early desktop calculators.[30] Texas Instruments also ceased its liquid crystal watch production during his tenure. In taking these steps, Heilmeier and his fellow TI executives were effectively acknowledging the existence of a highly competitive and increasingly international display industry.

• • •

The consumer electronics market had undergone a profound metamorphosis in the twenty-five years since David Sarnoff first envisioned a wall-mounted television based on light amplification. In 1951, the United States retained a commanding position in the global economy and an abiding faith in the linear model. By the mid-1970s, long-term fundamental research programs were falling out of fashion, and postwar reconstruction efforts had given rise to a new set of global competitors.[31] Capitalizing on the willingness of American firms to license their patents, companies in Europe and Asia remade themselves, first as regional leaders in electronics manufacturing and later as full-fledged rivals.[32] When RCA and other American corporations ceded their first-mover advantage in LCD production, businesses located elsewhere stepped in to fill the void.

One may partially attribute the success of these international LCD efforts to poor strategic planning among US firms—especially RCA—or foreign government policies that nurtured high-tech industries. Adopting either of these positions minimizes the indispensable part played by scientists and engineers throughout the technology-transfer process. Just as DSRC researchers took steps to ensure the circulation of information about liquid crystal materials and display assembly, their overseas counterparts built on their ideas and persuaded managers and government officials to sponsor flat-panel development.

Members of RCA's technical staff had engaged in transnational conversations about liquid crystals for several years before revealing their prototype displays. The "company-private" status of the LCD program did not prevent members of the DSRC team, including Heilmeier and Castellano, from attending the first International Liquid Crystal Conference at Kent State. Compared with future meetings, this 1965 gathering included relatively few guests from outside the United States; of the 128 participants, only fifteen Europeans were in attendance alongside a pair of scientists from India.[33] Nevertheless, at this event RCA liquid crystal researchers became acquainted with several people central to the subsequent growth of the LCD industry.

Foremost among these was British chemist George Gray. Gray had grown fascinated with liquid crystals while pursuing a doctorate at the University of London and later became a chemistry professor at the University of Hull. He recalled that it was difficult to attract support—or graduate students—to conduct investigations into liquid crystals, which in the 1950s were "seemingly without importance or technical value."[34] Before setting aside this work, he decided to write a book summarizing the state of the field. This 1962 publication was the first English-language monograph on liquid crystals, and within a few years it became a touchstone among the growing community of researchers interested in the subject.[35]

Now seen as a leading authority in a burgeoning research area, Gray became a regular speaker at liquid crystal workshops and conferences. It was at one such event in October 1968, arranged by the United Kingdom Ministry of Defence, that Gray met Cyril Hilsum, a semiconductor physicist from the Royal Radar Establishment (RRE). Hilsum oversaw RRE's flat-panel working group, which had been initiated the previous year when Her Majesty's Government realized that the licensing fees British firms paid RCA for its color TV picture tubes exceeded the development costs of the Concorde jet. He now approached Gray about the possibility of using liquid crystal displays to emancipate Britain from RCA's patent regime.[36]

Hilsum and Gray's interactions were quite positive, but the RRE working group was hesitant to embrace LCD technology, even after reviewing RCA's publications on dynamic scattering. It was not until December 1969 that they changed their minds, informing their superiors in a technical report that "one system is worth immediate attention. This is the display based on liquid crystals."[37] Gray was among the first scientists recruited to the project. He had remained in contact with Heilmeier's team, which was "opening the door and people's eyes to the possibility of display applications of LCs [liquid crystals]."[38] Now he was given a two-year military contract to establish a laboratory at Hull to examine "Substances Exhibiting Liquid Crystal States at Room Temperature" and lay the foundation for a broader British LCD research program.[39]

The Hull liquid crystal group, consisting of Gray and a lone postdoctoral fellow, set to work in the spring of 1970 seeking compounds that could replace the proprietary mixtures used in RCA's dynamic scattering displays. "We examined stilbenes, chlorostilbenes, and a range of esters including heterocyclic systems," Gray wrote later, "all with rather depressing results."[40] The Hull team was eager for a change in direction and found one in 1971 when they read Wolfgang Helfrich and Martin Schadt's *Applied Physics Letters* article on twisted nematic LCDs. Helfrich (another agent of liquid crystal technology transfer previously affiliated with RCA) and Schadt's displays required very different materials from Heilmeier's dynamic scattering devices. Following a brief internal debate, Gray received approval to recruit additional scientists to explore the newer approach.

The first few months of the Hull group's twisted nematic research proved just as disheartening as their previous efforts. Looking over their failures, Gray traced the problem to an unstable central linkage in the liquid crystals' molecular structure. Eliminating this instability became the team's foremost priority, and by late 1972 they had synthesized a new class of compounds known as cyanobiphenyls. Like RCA's liquid crystal group, the British chemists determined that mixtures of these compounds possessed a broader mesomorphic range. They soon found several combinations that were colorless, extremely stable at room temperature, and readily incorporated into twisted nematic displays. The commercial value of these new cyanobiphenyl materials was immediately apparent to everyone in the RRE project, but moving from small batches to bulk production would necessitate assistance from an experienced chemical manufacturer. On Hilsum's recommendation, the group reached out to British Drug House (BDH), which agreed to a partnership in December 1972.[41]

BDH set to work but found itself overwhelmed by requests for cyanobiphenyl samples following the publication of the Hull group's research in March 1973. Further complicating matters, that fall Glaxo—BDH's owner—decided to sell the smaller firm to Merck, the main German manufacturer of liquid crystal materials.[42] Some worried that Merck might wish to transfer all of BDH's production to Germany, but the new owners were wary of disrupting the solid working relationships between BDH, Hull, and RRE.[43] BDH therefore retained its autonomy and was able to announce a full line of "Exciting New Liquid Crystals" for sale by the end of the year.[44]

The preliminary excitement about BDH's liquid crystals took time to translate into high sales figures. Many LCD manufacturers had invested heavily in their own proprietary mixtures and were reluctant to abandon them. Consequently, BDH only sold approximately 150 grams of cyanobiphenyls to American companies during their first six months on the market. Fortunately for the British, by 1974, the LCD industry's center of gravity had drifted away from the United States.[45] While Gray and his colleagues continued to perfect their liquid crystal mixtures, BDH's leaders made inroads with a new group of customers who soon transformed their firm into the world's largest supplier of liquid crystal materials.[46] The Americans may have dismissed cyanobiphenyls, but BDH encountered a far more receptive audience in Japan.

• • •

RCA had cast a long shadow over the Japanese electronics industry since the early 1950s, when it offered firms such as Kōbe Kōgyō, Hitachi, Sony, and Tokyo Shibaura (later renamed Toshiba) access to its radio and television patents. The pace of RCA's international licensing activities quickened in the aftermath of the federal government's antitrust investigation. The 1958 consent decree that prevented RCA from collecting royalties on "radio-purpose" electronics applied only to domestic businesses, so the company embarked on an aggressive search for foreign clientele.[47] By 1960, Japan had grown into RCA's top market for patent licenses. That October, in recognition of his assistance jump-starting the Japanese economy, David Sarnoff traveled to Tokyo, where Emperor Hirohito awarded him the Order of the Rising Sun (third class): the highest honor ever bestowed on a foreign businessman.[48]

Six months after Sarnoff's trip, RCA opened a new laboratory in Tokyo. Intended to provide a link between the American and Japanese scientific communities, the new facility recruited a staff of Japanese scientists to conduct fundamental research into solid-state phenomena.[49] The laboratory coordinated a variety of "electrochemistry studies for displays," including work on color TV phosphors and electroluminescence, but liquid crystals never became a topic of major interest.[50] RCA encouraged personnel to meet with researchers from nearby universities and corporate laboratories, but these in-person interactions proved less important to the formation of Japan's LCD industry than the publicity blitz surrounding the DSRC group's 1968 press conference.[51]

Journalists from other countries were quick to join their American peers in spreading the word about RCA's latest breakthrough. Japanese media coverage was particularly extensive. Photographs of the DSRC research group appeared in newspapers, and NHK—the nation's public broadcasting network—aired the story repeatedly on the evening news.[52] Before the end of the year, an NHK film crew visited Princeton and interviewed Heilmeier about dynamic scattering for a documentary entitled "Firms of the World: Modern Alchemy."[53] "This research is expected to have various applications in the future," the film's narrator explained: "Making the flat-panel television with the liquid crystal would be quite an achievement. Thus, Japanese manufacturers are paying close attention to this."[54]

Sure enough, almost immediately after RCA's announcement, researchers across Japan commenced their own liquid crystal investigations. For some, like Yoshio Yamazaki, merely seeing dynamic scattering in action was sufficient to spark a preoccupation with LCDs. Yamazaki, a chemical engineer at Seiko, had been tasked with creating lubricants and plating techniques for mechanical watches until he saw a picture of RCA's liquid crystal clock in the *Asahi Shimbun* newspaper.[55] He swiftly persuaded his bosses to fund the development of an LCD watch, with the understanding that he would be held personally responsible should the project fail.[56] Yamazaki began by constructing dynamic scattering displays, but the lower power and high contrast of twisted nematic LCDs soon became more enticing. Seiko proceeded to negotiate a licensing agreement with Hoffman–La Roche and in September 1973 introduced its first LCD wristwatch.

Yamazaki's brainchild sparked a craze for digital watches in Japan as well as a spike in demand for better liquid crystal materials like BDH's cyanobiphenyls.[57]

Beyond the revelation that liquid crystals could be used in displays, RCA could claim little involvement in Seiko's digital watch venture. The same could not be said of other Japanese liquid crystal projects. Take the case of Tomio Wada. Like Yamazaki, Wada studied chemical engineering before obtaining a job at the Hayakawa Electric Company, one of Japan's oldest radio and television producers. Wada's supervisor, Tadashi Sasaki, had driven Hayakawa to enter the desktop calculator market and now mulled plans to create a handheld, battery-powered model assuming he could find a suitable low-power display.[58] Wada had hoped to solve the problem using electroluminescent phosphors but changed his mind in 1969 when he saw footage of NHK's visit to the DSRC. He recommended that Sasaki contact RCA to learn more about liquid crystals and determine whether they might permit the construction of a practical pocket calculator.[59]

Sasaki traveled to the United States to meet with members of RCA's technical staff and invite them to collaborate on the calculator project. In fact, Heilmeier and his colleagues had already contemplated such an application and, as a proof of concept, Louis Zanoni replaced the display in one of Hayakawa's desktop calculators with a dynamic scattering readout (fig. 6.1). Scientists in Princeton might have been sympathetic to Sasaki's proposal, but he found little support at RCA's semiconductor division in Somerville. Sasaki described his conversation with Bernard Vonderschmitt—one of the division's top managers—in a 1994 oral history. "At RCA, Vonderschmitt thought that the liquid crystal display is too costly," Sasaki explained. "At the time it was in an initial stage of production, so everything was expensive. Vonderschmitt thought that that kind of display should be used for industrial versions, but not for consumer goods."[60] (Fittingly, Vonderschmitt would later sign the contract authorizing the 1976 sale of RCA's liquid crystal operation to Timex.[61])

Sasaki may have left New Jersey broken hearted, but not empty handed. While RCA executives denied his partnership offer, they

Figure 6.1. Louis Zanoni retrofit this calculator with a dynamic scattering readout to demonstrate another possible application for LCD technology. RCA managers later rejected an offer to partner with Sharp Corporation, which ultimately commercialized the concept. (David Sarnoff Library Collection, courtesy of Hagley Museum and Library.)

were more than willing to supply his firm with a license to produce dynamic scattering displays. In a later interview, Sasaki recalled that Hayakawa Electric, renamed Sharp Corporation in 1970, paid approximately $3 million for the patent rights.[62] A high price perhaps, but a necessary down payment if Sharp wished to move forward with the LCD calculator on its own. To that end, the company organized a classified development effort code-named Project S734; the S stood for "secret" and 734 referred to the company's intended deadline—April 1973 (or 1973-04).[63] Wada, who had first suggested the idea that Sharp should invest in liquid crystals, now oversaw an interdisciplinary R & D group that raced to master all aspects of LCD fabrication. The learning curve was steep, but the Sharp team resolved all of the chemical and electrical engineering issues associated with the construction of a dynamic scattering calculator in the spring of 1973, weeks before the planned completion date.[64]

That May, Sharp released the Elsi Mate EL-805, the first portable calculator with a liquid crystal readout. Originally intended for the business sector, the EL-805 soon became popular in households, classrooms, and even outdoor venues such as construction sites. The slow response time of its dynamic scattering display in cold temperatures led some to refer to it as the "Obake" ("ghost") calculator, but the profits from EL-805 allowed Sharp to expand its liquid crystal factories. The firm finally exorcised the ghost by switching to twisted nematic displays based on BDH's cyanobiphenyl mixtures.[65] By the end of the 1970s, Sharp had invested a total of $200 million on LCD manufacturing, solidifying its position as a leading flat-panel display producer.[66] Only in hindsight did it become clear to RCA's leaders that in turning down Sasaki's proposal, they had squandered one of their last opportunities to gain a foothold in the Japanese electronics market.

No technology embodied the widening technological gap between East and West as much as the LCD television, which finally reached consumers in the 1980s thanks to Japanese scientists and engineers. In 1983, a decade after a DSRC task force rejected liquid crystals as the basis for a flat-panel television, engineers from Seiko demonstrated a wristwatch-size TV that would put Dick Tracy's to shame. Each of its twisted nematic pixels served as a high-speed shutter that controlled the passage of white light. The inclusion of red, green, and blue filters enabled the presentation of full-color images.[67] Every pixel was activated by a corresponding transistor, an approach that RCA's Bernard Lechner had demonstrated with discrete components in 1968. The Seiko project drew on Lechner's ideas as well as those of Westinghouse physicist T. Peter Brody, who had constructed his own LCD addressed with an integrated array of thin-film transistors in

1973.[68] (Brody would later coin the term "active matrix" to refer to this setup.[69]) Seiko's two-inch diagonal screen would never replace a home TV set, but its use of thin-film transistors—another RCA invention—to produce color pictures established the operating principles for all future LCD televisions and contradicted the conventional wisdom that liquid crystals were only useful in simple applications such as watches or calculators.[70]

Sharp had been among the doubters. No matter how much money their LCD calculators earned, the company's research staff felt that electroluminescent phosphors were a more promising basis for a flat-panel TV. Seiko had proven them wrong, and although Sharp was able to cobble together a three-inch diagonal LCD television by 1987, a bold response would be necessary to regain leadership in the field.[71] Isamu Washizuka, the general manager of Sharp's Liquid Crystal Division, consulted with company engineers, who pushed for an incremental development program building progressively larger displays over the course of several years. Washizuka instead proposed a radical leap in screen size, toward something comparable with a standard CRT television. It was a plan whose audacity resembled Sharp's calculator initiative, perhaps deliberately so, since Washizuka had been instrumental in setting up Project S734. Once again, Sharp's engineers rose to the challenge, pushing the limits of existing manufacturing equipment to build a fourteen-inch color LCD screen in under a year (fig. 6.2).[72]

Sharp presented its handiwork to the public in June 1988. The announcement marked the beginning of one story and the end of another. On the one hand, Washizuka and his team showed that liquid crystals could match the size and performance of a cathode-ray tube, effectively launching the modern LCD television industry. On the other, Sharp had at long last fulfilled David Sarnoff's prediction of full-color TV sets that could hang like pictures on a wall. One can only imagine the General's disappointment with his subordinates, who vetoed an alliance with the firm responsible for this feat nearly twenty years earlier.

Figure 6.2. Isamu Washizuka, general manager of Sharp Corporation's Liquid Crystal Division, posing with his company's 14-inch color LCD television in 1988. (Hirohisa Kawamoto, "The History of Liquid Crystal Displays," *Proceedings of the IEEE* 90, no. 4 [Apr. 2002], 495. Reprinted with permission from IEEE and Sharp Corporation.)

• • •

Sharp's fourteen-inch television may have been the first major salvo in the electronics industry's revolt against the cathode-ray tube, but it would take another two decades to topple the older technology. Washizuka's prototype was too expensive for the average consumer, and as the LCD industry struggled to reduce costs, it faced competition from challengers who had embraced an alternative display technology: the plasma panel. Plasma displays used thousands of miniature gas discharge tubes, which operated much like fluorescent light bulbs, to generate a picture. Early plasma displays could only create monochrome images, but by 1995 the Japanese firm Fujitsu was producing full-color plasma television sets with a forty-two-inch diagonal. Plasma screens had several drawbacks compared to LCDs. They were heavier, used more power, and relied on phosphors that degraded over time, but they could be easily scaled up in size and offered excellent picture quality. For many, plasma displays' strengths outweighed their weaknesses, and plasma panels captured a large portion of flat-panel television market until the mid-2000s.[73]

Compared with plasma displays, liquid crystals took a more circuitous path to people's living rooms. Once again, the key to their success was their portability. LCDs' low power requirements made them ideal for use in handheld video games, mobile phones, and most importantly, laptop computers. By the mid-1990s, notebook computer screens constituted 80 percent of the color active matrix LCD market.[74] This increased demand drove LCD manufacturers to ramp up production and devise more sophisticated fabrication procedures. Screen sizes expanded and the price ratio between a liquid crystal television and a comparably sized CRT dropped from 10:1 at the start of the decade to 5:1 in 1997.[75] The arrival of HDTV and the popularity of high-resolution DVD movies reinforced these trends, and at the end of 2007, LCD televisions began outselling both plasma panels and CRTs.[76]

Thanks to companies like Sharp and Seiko, Japanese businesses accounted for 95 percent of LCD production during the 1990s.[77] Some of those firms, such as Sony and Panasonic, remain important players in the twenty-first-century LCD industry, but Japan's display dominance has dwindled significantly in the face of competition from South Korea, China, and Taiwan. In 2016, for example, firms in South Korea and China were responsible for nearly three-quarters of the LCD television market compared with Japan's 16 percent share.[78] Far removed as these up-and-coming East Asian enterprises might seem from Princeton, they also benefitted from the insights of DSRC liquid crystal researchers.

In some cases, the RCA connection was somewhat distant. Several leading Taiwanese electronics companies, including AU Optronics and Chimei Optoelectronics, were introduced to LCD production through partnerships with IBM, whose staff first heard about the technology at licensing seminars hosted by Heilmeier's team in the 1960s and 1970s.[79] But at times, former members of RCA's technical staff were direct participants in East Asian liquid crystal operations. Some acted independently, like Nunzio Luce, who joined with fellow Optel alumnus Joel Goldmacher to found Springwood Electronics, a Hong Kong–based producer of LCD watches.[80] Others followed Castellano's example and served as consultants, providing guidance to managers unfamiliar with the logistics of LCD fabrication. When, on one noteworthy occasion in 1984, a South Korean TV manufacturer became interested in replacing its CRTs with flat-panel displays, they contacted one of their countrymen: former RCA chemist Chan Soo Oh. Although still employed at Beckman Instruments, Oh obtained permission to travel to the suburbs of Seoul, lead workshops on liquid crystal chemistry, and assist in the selection of factory equipment. He was among the earliest experts who helped his client—Samsung Electronics—accumulate the necessary knowledge base to assemble its own LCDs.[81] Samsung has since grown into the world's largest flat-panel display producer, whose screens can be found in one out of every four LCD televisions sold today.[82]

Korea may now be "the center of the liquid-crystal universe," as Nicholson Baker argued in a 2013 *New Yorker* article, but if so, it is only the latest location to claim that title.[83] Gaze at the constellation of screens that surrounds us each day, and regardless of their provenance, one can detect traces of the LCD's origins stretching inexorably back toward New Jersey. Every pixel on your television, laptop, and cell phone bears the imprint of the RCA personnel who took a far-fetched technological dream—a television that could hang on the wall—and turned it into something real. Behind each tiny dot is a liquid crystal mixture, woven into a molecular helix, and coaxed into allowing light to reach your eye with an intricate web of circuitry. All of these innovations—room-temperature mesomorphic materials, the twisted nematic operating mode, and active matrix addressing—emerged from RCA's laboratories.

In a sense, these displays are far more powerful tributes to RCA's technical staff and the thousands of others who refined and improved on their work over the past half century than any commemorative plaque. Traditional monuments are bounded in space and time, but the influence of RCA researchers was neither confined to the DSRC nor to their company's brief foray into LCD manufacturing. They were the nucleus of a knowledge community, one that formed in response to the technical challenge of creating a flat-panel replacement for the CRT.[84] As practicing scientists and engineers, aspiring entrepreneurs, and industry consultants, they transmitted information about liquid crystal materials and applications around the world and were often pivotal to their commercialization. The results of their participation were rarely evident to the average consumer, but the screens in our pockets, homes, and workspaces, so common today as to become practically invisible, are tangible evidence of their ingenuity.

On any given day, millions of people will check the time, peruse their e-mail, share a photograph, or stream a favorite movie on an LCD without knowing the name of anyone affiliated with RCA's liquid crystal operation. Yet despite this obscurity, their legacy is secure. There were flat-panel displays before the LCD, but the interconnected, digital society in which we now live depends on low-power, portable displays descended from the prototypes George Heilmeier and his team demonstrated in 1968. Another technology, such as the increasingly prominent organic light-emitting diode (OLED), may someday displace liquid crystals in many consumer applications, but if so we will just be substituting one set of flat screens for another. Whether the political, economic, and cultural effects of that transition will prove as significant as the introduction of the LCD, or any other electronic display, remains to be seen. Understanding the origins of these technologies and the parts that corporate scientists and research managers played in their creation is a necessary first step toward answering these questions. Only in this way can we hope to peer beyond the surface of our screens and catch a glimpse of the world that they have made.

ACKNOWLEDGMENTS

On April 10, 1968, the month before RCA announced the creation of the first liquid crystal displays, Alfred Hitchcock received his only Academy Award. Though his film *Rebecca* was named Best Picture in 1940, producer David O. Selznick had accepted the prize, leaving Hitchcock without an Oscar to call his own. Now, nearly three decades later, the Academy's board of governors decided to recognize the filmmaker's contributions to modern cinema with the Irving G. Thalberg Memorial Award for lifetime achievement. Following a summary of his career by fellow director Robert Wise, Hitchcock strolled across the stage to the familiar strains of Gounod's "Funeral March of a Marionette" and delivered one of the shortest Oscar speeches on record: "Thank you."

Hitchcock's example came to mind as I started to compile a list of the many people who assisted with the creation of this book and realized there was no way that I could adequately express my gratitude to everyone involved. The elegance of the director's solution to this problem was understandably appealing, but embracing this tactic would run counter to my efforts to restore complexity to accounts of collaborative research. While some might revel in the irony of such a stylistic maneuver, in the following pages I will instead attempt to call attention to the most prominent contributors to this project's success.

Like the LCD itself, this book began in Princeton. I am grateful to Alexander Magoun, the longtime director of the David Sarnoff

Library, who first suggested that RCA's liquid crystal program might be a worthwhile research topic. In addition to sharing his encyclopedic knowledge of RCA and its history, Alex granted me unprecedented access to the firm's technical archive, ensuring that I accumulated enough material to sustain my work after the Sarnoff Library's closure in 2009. Special thanks are similarly due to the Program in History of Science at Princeton University, especially Angela Creager, Emily Thompson, and above all, Michael Gordin. From the beginning, Michael has proven a stalwart mentor, editor, and sounding board, and my scholarship has consistently benefitted from his insights. I would also be remiss if I did not acknowledge the late Michael Mahoney, whose graduate seminars provided my formal introduction to the history of technology.

During my time in Princeton, I made heavy use of the university's library system with help from Melissa Acosta, Elizabeth Bennett, Willow Dressel, Stephen Ferguson, and the staff of the Seeley G. Mudd Manuscript Library. I subsequently presented several early chapter drafts at the history of science program's weekly seminar. Among the participants in those discussions, I wish to thank Melinda Baldwin, Michael Barany, Dan Bouk, Howard Chiang, Henry Cowles, William Deringer, Yulia Frumer, Anthony Grafton, Evan Hepler-Smith, Robert MacGregor, Christopher McDonald, Susan Naquin, Margaret Schotte, Ksenia Tatarchenko, Iain Watts, and Adrian Young for their perceptive questions and unwillingness to settle for easy answers.

While the preliminary research for this book took place in Princeton, it draws heavily on other archival collections as well. Foremost among these is the Hagley Museum and Library, which previously housed the technical archive from RCA's Camden facilities and has since done an outstanding job processing the Sarnoff Library Collection under the leadership of Erik Rau and archivists Daniel Michelson and Kenneth Cleary. Thanks as well to Hagley audiovisual archivists Kevin Martin, Lynsey Sczechowicz, and Jon Williams for uncovering most of the images found in the preceding chapters. The Sarnoff Library's artifacts—including multiple flat-panel display prototypes—were sent to the College of New Jersey, and it was an honor working with Emily Croll to investigate them in connection with several exhibitions on RCA and the history of telecommunications.

Beyond these RCA-specific collections, I received further research assistance from M. E. Brennan at the Institute of Electrical and Electronics Engineers (IEEE); Robert Colburn, Michael Geselowitz, and Sheldon Hochheiser at the IEEE History Center; George Curley at the University of California, Riverside; Kim Feamster and Paul Gagnon at IHS Markit; Nicola Finn at Futuresource; Lou Galie, David Payne, and the legal department at Timex Corporation; Cara Gilgenbach at Kent State University's Special Collections and Archives; Darrel Hopper and Frederick Meyer at the Air Force Research Laboratory at Wright-Patterson Air Force Base; Jim Maxwell at the Liquid Crystal Institute; Toshiyuki Matsumura at Sharp Corporation; Rosemary Morrow at Redux Pictures; and Carlene Stephens at the National Museum of American History.

A major advantage of studying recent history is the opportunity to interact with people who either witnessed or were directly involved in the events under consideration. I am *extremely* grateful to the former members of RCA's technical staff who supplied oral histories in connection with this project and to Joseph Castellano, Zoltan Kiss, Richard Klein, Joseph Kleitman, Naomi Kleitman, Bernard Lechner, Robert Quinn, Greg Zanoni, and Louis Zanoni, whose personal collections of documents and photographs offered glimpses into day-to-day research activities at RCA and Optel. Thanks as well to the participants in the monthly RCA alumni luncheon in Princeton for sharing stories about the company's glory days while dining at Conte's Pizza, Olive Garden, or Super Star East Buffet.

The task of transforming my research into a book commenced at the Chemical Heritage Foundation (CHF), where I was privileged to receive fellowships from the Beckman Center for the History of Chemistry and the Institute for Research. Many thanks to Ronald Brashear, Carin Berkowitz, and Jody Roberts for granting me membership in such a vibrant community of scholars. Not only did CHF provide me with the time and resources necessary to draft my manuscript, but I obtained valuable feedback through the organization's writing group and brown-bag lecture series.

Outside of these events, I enjoyed productive conversations with CHF staff members and research fellows, including Lee Berry, Donna Bilak, Jacqueline Boytim, Clay Cansler, David Caruso, Deanna Day, Hilary Domush, Elisabeth Berry Drago, Matthew Eisler, Monica Fonorow, Neil Gussman, David Haldeman, Nicholas Harris, Sarah Hunter-Lascoskie, Hillary Kativa, Bob Kenworthy, Joel Klein, Andrew Mangravite, Michal Meyer, Rebecca Ortenberg, Alexis Pedrick, Evan Ragland, Sarah Reisert, Kathy Seufert, Patrick Shea, Matthew Shindell, Preston Stone, Anke Timmermann, Brigitte van Tiggelen, James Voelkel, Jessica Wade, and Nasser Zakariya.

While at CHF, I also regularly attended the annual meetings of the Society for the History of Technology and the History of Science Society. These conferences brought me into contact with several scholars whose work further informed this book's arguments, most notably David Brock, Hyungsub Choi, Jonathan Coopersmith, Margaret Graham, David Hounshell, Paul Israel, Ann Johnson, Christophe Lécuyer, Stuart Leslie, Patrick McCray, Cyrus Mody, Adelheid Voskuhl, Matthew Wisnioski, and Audra Wolfe.

Karen Merikangas Darling, Evan White, and the rest of the staff at the University of Chicago Press have done a wonderful job shepherding my book toward publication. I also appreciate the editorial suggestions of the three anonymous readers who reviewed my manuscript. Final revisions to the text occurred after I relocated to Kansas City to take up a new position at the Linda Hall Library, and I would like to thank Lisa Browar, Michelle Lahey, Tania Munz, and my colleagues in the Research & Scholarship department for their support as this project draws to a close. Thanks also to Ben Gibson and Jon Rollins from the Library's Digital Initiatives Unit for scanning several of the images featured in this volume.

Scholarship can be a lonely endeavor, particularly when one has recently moved to a new city, but over the years I have drawn strength from friendships forged as a student in Connecticut, a science teacher in Philadelphia, and a historian turned occasional quizmaster in New Jersey. Though we have long since scattered across the globe, I am fortunate to remain in touch with Mike Anderson, Jeremy Banzhaf, Liz and Keith Bonawitz, Evan Burke, Miriam Clinton, Evan Davidson, Jesse Eshkol, Mike Gann, Colin and Julie Girgenti, Emily Goodwin, Mac and Beth Haas, Alex Kloth, Jenna Koch, J. D. Martin, Scott Meek, Matthew Mugmon, John Murphy, Christophe Renaud, Alex Rosati, James Sinclair, Sean and Sarah Tulin, James Turner, Charles Umiker, the entire Woodruff clan (especially David, Matthew, and Kate), and Afton Woodward.

In the end, I owe the deepest debt of all to my family. When I was in elementary school, my father Jeffrey introduced me to RCA through a screening of Ken Burns's *Empire of the Air*, little realizing that it would inspire a lifelong fascination with the history of electronics. From my mother Janna, I inherited my love of libraries, which set the course for my current career path. Although separated by hundreds of miles, I have always been able to count on my sister Andrea's optimism and humor to cheer me up after a long day of research or writing. I am grateful to everyone who helped me tell this story, but it is to my family that this book is ultimately dedicated.

Thank you.

NOTES

Introduction

1. RCA News, "RCA Announces Breakthrough in Liquid Crystal Field," press release, 28 May 1968, p. 1 David Sarnoff Library Collection, Hagley Museum and Library, Wilmington, DE (DSL).

2. RCA Department of News and Information, "Fact Sheet: Liquid Crystal Displays," [28 May 1968], p. 1, Louis Zanoni Personal Archive, Ewing, NJ (ZPA).

3. James Hillier, "Draft of Remarks by Dr. Hillier at Liquid Crystal Press Conference, 28 May 1968," p. 2, Bernard Lechner Collection, DSL.

4. Ibid., 2–3.

5. Ibid., 4–6.

6. Ibid., 7.

7. Ibid., 8.

8. The exact size of the LCD industry is difficult to determine because of the technology's presence in so many different applications (TVs, laptops, etc.). To provide a sense of its scale, however, IHS Markit—a leading source of information about the consumer electronics sector—estimates that in 2016, the value of the LCD television market was $83.4 billion, corresponding to approximately 223 million total units shipped. Futuresource, another consulting firm that collects data on the display industry, projects that in 2017, the number of LCD television shipments will increase to 225.7 million units. This information was derived from Nicola Finn (Futuresource), e-mail message to author, 6 Feb. 2017, and Paul Gagnon (senior manager of analysis and research, IHS Markit), telephone conversation with author, 2 Feb. 2017.

9. For more on the origins and proliferation of the linear model, see Glen Ross Asner, "The Cold War and American Industrial Research" (PhD diss., Carnegie Mellon University, 2006); David Hounshell, "The Evolution of Industrial Research in the United States," in *Engines of Innovation: U.S. Industrial Research at the End of an Era*,

ed. Richard S. Rosenbloom and William J. Spencer (Boston: Harvard Business School Press, 1996), 13–85; Benoit Godin, "The Linear Model of Innovation: The Historical Construction of an Analytical Framework," *Science, Technology & Human Values* 31, no. 6 (Nov. 2006): 639–67.

10. Clayton M. Christensen, *The Innovator's Dilemma* (New York: Harper Business, 2011).

11. Hyungsub Choi, "The Boundaries of Research: Making Transistors at RCA, 1948–1960," *Technology and Culture* 48, no. 4 (Oct. 2007): 780.

12. See, e.g., Sally H. Clarke, Naomi R. Lamoreaux, and Steven W. Usselman, eds., *The Challenge of Remaining Innovative: Insights from Twentieth-Century American Business* (Palo Alto. CA: Stanford Business Books, 2009); David Hounshell and John Kenly Smith Jr., *Science and Corporate Strategy: Du Pont R&D, 1902–1980* (New York: Cambridge University Press, 1988); Emerson Pugh, *Building IBM: Shaping an Industry and Its Technology* (Cambridge, MA: MIT Press, 1995).

13. Hounshell, "Evolution of Industrial Research." See also Thomas P. Hughes, *American Genesis: A Century of Invention and Technological Enthusiasm, 1870–1970* (1989; repr., Chicago: University of Chicago Press, 2004), chap. 4.

14. Steven Shapin, *The Scientific Life: A Moral History of a Late Modern Vocation* (Chicago: University of Chicago Press, 2008), 127–32.

15. R. Joseph Anderson and Orville R. Butler, *History of Physicists in Industry* (College Park, MD: American Institute of Physics, 2008), https://www.aip.org/history/pubs/HOPI_Final_report.pdf.

16. Shapin, *Scientific Life*, 114 (emphasis in original).

17. Ibid. See also Jan Golinski, *Making Natural Knowledge: Constructivism and the History of Science*, 2nd ed. (Chicago: University of Chicago Press, 2005), chap. 2.

18. National Science Board, *Science and Engineering Indicators 2016*, NSB-2016-1 (Arlington, VA: National Science Foundation, 2016), 3–37. In total, 70.1 percent of American scientists and engineers worked in the business sector, with 52.4 percent in for-profit businesses, 11.1 percent in nonprofit organizations, and 6.6 percent self-employed.

19. Jefferson Cowie, *Capital Moves: RCA's Seventy-Year Quest for Cheap Labor* (Ithaca, NY: Cornell University Press, 1999).

20. David Sarnoff, "Electronic Revolution, Present and Future," *New York Times Magazine*, 30 Sept. 1956, 38.

21. Fred Aftalion, *A History of the International Chemical Industry: From the "Early Days" to 2000*, 2nd ed. (Philadelphia: Chemical Heritage Foundation Press, 2001), chap. 5. See also Hounshell and Smith, *Science and Corporate Strategy*, 327–501.

22. Aftalion, *International Chemical Industry*; Michael Riordan and Lillian Hoddeson, *Crystal Fire: The Birth of the Information Age* (New York: W. W. Norton, 1997). For more on RCA specifically, see R. W. Peter, "Materials Research," *RCA Engineer* 5, no. 4 (Dec. 1959–Jan.1960): 4–9; Bruce H. Shore, *The New Electronics* (New York: McGraw Hill, 1970).

23. For more on the centrality of materials to the history of electronics, see David C. Brock and Christophe Lécuyer, "Digital Foundations: The Making of Silicon-Gate Manufacturing Technology," *Technology and Culture* 53, no. 3 (July 2012): 561–97; Christophe Lécuyer and David C. Brock, "The Materiality of Microelectronics," *History and Technology* 22, no. 3 (Sept. 2006): 301–25; Christophe Lécuyer and Takahiro Ueyama, "The Logics of Materials Innovation: The Case of Gallium Nitride and Blue Light Emitting Diodes," *Historical Studies in the Natural Sciences* 43, no. 3 (June 2013): 243–80.

24. For more on the interplay between managers and engineers throughout the design process, see Ann Johnson, *Hitting the Brakes: Engineering Design and the Production of Knowledge* (Durham, NC: Duke University Press, 2009), 3–13.

25. Al Pinsky, "Preliminary Plans for Press Conference to Announce RCA Development of Liquid Crystal Electron Optic [*sic*] Effect," interdepartmental correspondence, 10 Apr. 1968, p. 2, Bernard Lechner Collection, DSL.

26. RCA Department of News and Information, "Liquid Crystal Displays."

27. Hillier, "Draft of Remarks"; George Heilmeier, "Press Conference Presentation," [28 May 1968], Bernard Lechner Collection, DSL.

28. Joseph Castellano lists more than thirty newspapers and magazines that reported on the LCD press conference in *Liquid Gold: The Story of Liquid Crystal Displays and the Creation of an Industry* (Hackensack, NJ: World Scientific, 2005), 55. See, e.g., William K. Stevens, "Display Devices Crystallize for R.C.A.," *New York Times*, 29 May 1968, 47, 54; "New Thin-Screen Displays Utilize Liquid Crystal Effect," *IEEE Spectrum* 5, no. 7 (July 1968): 129; "RCA Develops a New Visual Display Means Using Liquid Crystals," *Wall Street Journal*, 29 May 1968, 12.

29. Stevens, "Display Devices"; "In TV Future," *Washington Post*, 29 May 1968, D9.

30. For more on genres of scientific communication, see Peter Dear, ed., *The Literary Structure of Scientific Argument: Historical Studies* (Philadelphia: University of Pennsylvania Press, 1991); Frederic L. Holmes, Jürgen Renn, and Hans-Jörg Rheinberger, eds., *Reworking the Bench: Research Notebooks in the History of Science* (Boston: Kluwer Academic, 2003); Greg Myers, "From Discovery to Invention: The Writing and Rewriting of Two Patents," *Social Studies of Science* 25, no. 1 (Feb. 1995): 57–105.

Chapter One

1. *Dedication of Plaque to David Sarnoff: Commemorating His 45 Years of Service in Radio and Naming the Princeton Laboratories of RCA as the David Sarnoff Research Center* (Princeton, NJ: RCA Laboratories Division, 1951), 5, 14. For examples of press coverage, see "Sarnoff Honored; 45 Years in Radio," *New York Times*, 28 Sept. 1951, 32; "Electrical Voltages Derived by RCA from Radioactive Materials," *Wall Street Journal*, 28 Sept. 1951, 16.

2. For more on RCA's origins, see Margaret B. W. Graham, *The Business of Research:*

RCA and the VideoDisc (New York: Cambridge University Press, 1986), chap. 2; Robert Sobel, *RCA* (New York: Stein and Day, 1986), chap. 1–2; J. C. Warner, "A History of Radio Corporation of America: The Years to 1938," *RCA Engineer* 3, no. 1 (June–July 1957): 3–9.

3. Sarnoff's career is detailed in several biographies, including Kenneth Bilby, *The General: David Sarnoff and the Rise of the Communications Industry* (New York: Harper and Row, 1986); Carl Dreher, *Sarnoff: An American Success* (New York: Quadrangle/New York Times Book Co., 1977); and Eugene Lyons, *David Sarnoff: A Biography* (New York: Harper and Row, 1966).

4. Graham, *Business of Research*, 42–47.

5. Kenyon Kilbon, "Pioneering in Electronics: A Short History of the Origins and Growth of RCA Laboratories, Radio Corporation of America, 1919 to 1964" (unpublished typescript, revised Aug. 1964), chap. 5. Paginated copy available at http://www.davidsarnoff.org/kil.html.

6. *Dedication*, 12–13.

7. In addition to the articles cited in note 1, see "Inventions Wanted," *Time*, 8 Oct. 1951, 54; "General David Sarnoff, Miracle Man of the Electronic Age," *Advertiser*, Oct. 1951, 13, 42. For an internal perspective, see "Sarnoff Challenges Scientists of RCA to Make Three Important Inventions," *Radio Age* 11, no. 1 (Oct. 1951): 5–7, 31. *Radio Age* was RCA's in-house magazine, published quarterly by the firm's Department of Information.

8. *Dedication*, 18.

9. Ibid., 22–23.

10. Ibid., 15–27; Kilbon, "Pioneering in Electronics," 198–200; Lyons, *David Sarnoff*, 300–302.

11. Bilby, *The General*, 5–6.

12. *Dedication*, 18–19.

13. Ibid., 19.

14. Ibid.

15. Ibid.

16. George H. Brown, *And Part of Which I Was: Recollections of a Research Engineer* (Princeton, NJ: Angus Cupar, 1982), 268.

17. Graham, 54.

18. Brown, 277–78; Graham, *Business of Research*, 67n16. William Webster, who assumed leadership of the David Sarnoff Research Center in 1968, agreed with his predecessors, suggesting that "the guys that put the ideas into that Sarnoff speech were the management of the labs at the time" in a 21 Apr. 2010 telephone conversation with the author.

19. For a summary of RCA's wartime research, see Kilbon, "Pioneering in Electronics," chap. 6.

20. Sobel, *RCA*, 161–62.

21. For more on this dispute, see Russell W. Burns, *The Struggle for Unity: Colour Television, the Formative Years* (London: Institution of Engineering and Technology, 2008), chap. 6–7; David Fisher and Marshall Jon Fisher, *Tube: The Invention of Television* (Washington, DC: Counterpoint, 1996), chap. 17; Sobel, *RCA*, chap. 8.

22. "TV Firm Plans Suit to Bar Color Move," *New York Times*, 12 Oct. 1950, 50.

23. "The General," *Time*, 23 July 1951, 74.

24. Kilbon, "Pioneering in Electronics," 181.

25. "Introducing Your Engineering Leadership," *RCA Engineer* 1, no. 1 (June–July 1955): 25.

26. RCA Laboratories, *Research Report 1951*, 16, David Sarnoff Library Collection, Hagley Museum and Library, Wilmington, DE (DSL); Graham, *Business of Research*, 70; Kilbon, "Pioneering in Electronics," 182–84.

27. In addition to the materials listed in the introduction (n. 9), see also Donald E. Stokes, *Pasteur's Quadrant: Basic Science and Technological Innovation* (Washington, DC: Brookings Institution, 1997), chap. 1–2.

28. Kilbon, "Pioneering in Electronics," 179.

29. RCA Laboratories, *Research Report 1951*, 24, DSL. The same report indicates that the DSRC's technical staff had 293 staff members (p. 15) as opposed to the 125 at its opening. For the latter figure, see RCA Laboratories, *1942–1967: Twenty-Five Years at RCA Laboratories* (Philadelphia: Smith-Edwards, 1967), 2.

30. RCA Laboratories, *Research Report 1951*, 67–68.

31. B. R. Crisler, "New York Sees Color Television on Theatre-Size Screen," *Christian Science Monitor*, 17 Oct. 1951, 7.

32. "Color Television on Theatre Size Screen," *Radio Age* 11, no. 1 (Oct. 1951): 3–4, 31; Kilbon, "Pioneering in Electronics," 78, 225.

33. Donald G. Fink, *Television Engineering*, 2nd ed. (New York: McGraw-Hill, 1952), 142–43.

34. RCA Laboratories, *Research Report 1954*, 29–30, DSL; Solomon Lasof and David Epstein, *Color Television Projectors for the Home*, PTR-509, 25 Oct. 1955, pp. 3–4, DSL.

35. RCA Laboratories, *Research Report 1955*, 75, DSL; Lasof and Epstein, *Color Television Projectors*, 15, 23–25.

36. "Nicoll, Frederick H.," in DSRC Biographies Collection, Box 4, Folder 15, DSL.

37. "Contributors," *Proceedings of the IRE* 45, no. 10 (Oct. 1957): 1423.

38. Nicoll and Kazan are listed as having won an award "for team performance in the conception and development of a triple-cathode gun for color kinescopes" in RCA Laboratories, *Research Report 1952*, 3.

39. L. E. Tannas Jr., "Electroluminescent Displays," in *Flat-Panel Displays and CRTs*, ed. Lawrence E. Tannas Jr. (New York: Van Nostrand Reinhold Company), 240.

40. Humboldt W. Leverenz, *An Introduction to the Luminescence of Solids* (New York: John Wiley and Sons, 1950), 152.

41. Watson Davis, "Electric Light without Bulbs," *Science News Letter* 59 no. 25 (23 June 1951): 394–95; "Lights from Panelescence," *Wall Street Journal*, 14 July 1953, 13.

42. Kilbon, "Pioneering in Electronics," 329–31. Leverenz later discounted efforts to assemble a flat-screen display, arguing that changing the form factor would not fundamentally alter the television viewing experience. See Humboldt W. Leverenz, an oral history conducted 15 July 1975 by Mark Heyer, IEEE History Center, Hoboken, NJ, http://ethw.org/Oral-History:Humboldt_W._Leverenz.

43. RCA Laboratories, *Research Report* 1952, 99.

44. Kilbon, "Pioneering in Electronics," 247–52.

45. RCA Laboratories, *Research Report* 1952, 99.

46. RCA Laboratories, *Research Report* 1953, 23.

47. B. Kazan and F. H. Nicoll, *An Electroluminescent Light-Amplifying Picture Panel*, PTR-458, 16 May 1955, Sheet I, 15–18, DSL; RCA Laboratories, *Research Report* 1954, 81–82.

48. Kazan and Nicoll, *Electroluminescent Light-Amplifying Picture Panel*, PTR-458, 39.

49. RCA Laboratories, *Research Report* 1953, 4.

50. *Dedication*, 27.

51. Ibid., 23.

52. Dr. Jan Rajchman, oral history interview conducted 11 July 1975 by Mark Heyer and Al Pinsky, IEEE History Center, Hoboken, NJ, http://ethw.org/Oral-History:Jan_Rajchman; Jan Rajchman, interview by Richard Mertz, 26 Oct. 1970. Computer Oral History Collection, 1969–1973, 1977, Archives Center, National Museum of American History, http://amhistory.si.edu/archives/AC0196_rajc701026.pdf.

53. For more on Rajchman's contributions to computer memory development, see Kilbon, "Pioneering in Electronics," 323–27; "The Scientist as Inventor and Manager: An Interview with Dr. J. A. Rajchman," *RCA Engineer* 19, no. 2 (Aug.–Sept. 1973): 48–49.

54. Jan Rajchman, File 334, p. 1, DSL.

55. Ibid., 1–20; Jan Rajchman, Notebook P-2519, pp. 38–55, DSL.

56. RCA Laboratories, *Research Report* 1952, 96–98.

57. RCA Laboratories, *Research Report* 1953, 91–94.

58. Rajchman, interview by Mark Heyer and Al Pinsky, 11 July 1975.

59. J. A. Rajchman and A. W. Lo, "The Transfluxor," *Proceedings of the IRE* 44, no. 3 (Mar. 1956): 321–32.

60. Jan A. Rajchman and A. W. Lo, *The Transfluxor*, PTR-473, 20 May 1955, pp. 2, 40, DSL.

61. Jan Rajchman, Notebook 41, pp. 1–10, 13–17, 20–25, 27–79, 90–97, DSL.

62. Jan A. Rajchman, *Mural Television*, PEM-1013C, 26 Oct. 1955, pp. 7–9, DSL.

63. George Briggs, interview by author, 12 Oct. 2009, Princeton, NJ.

64. Rajchman, Notebook 41, p. 52.

65. Rajchman, *Mural Television*, 9.

66. J. A. Rajchman, G. R. Briggs, and A. W. Lo, "Transfluxor Controlled Electroluminescent Display Panels," *Proceedings of the IRE* 46, no. 11 (Nov. 1958): 1819.

67. Rajchman, *Mural Television*, 10.

68. Briggs, interview, 12 Oct. 2009.

69. Ibid.

70. Rajchman, *Mural Television*, 10–11.

71. Ibid., 13.

72. Jan Aleksander Rajchman, Electrical Display Device, US Patent 2,928,894, filed 31 May 1955 and issued 15 Mar. 1960.

73. Ibid., 22.

74. RCA Laboratories, *Research Report 1955*, 2.

75. "Aspects of Broadcasting, Present and Future," *Radio Age* 13, no. 4 (Oct. 1954): 7.

76. W[aldemar]. K[aempffert]., "Bright Future for Electronics," *New York Times*, 24 Oct. 1954, E9.

77. "Aspects of Broadcasting," 7.

78. Elmer Engstrom, "Horizons in Electronics: A Revolution in Materials," WESCON, San Francisco, 26 Aug. 1955, p. 10. Elmer W. Engstrom Papers, Box 1, Vol. 8, Department of Rare Books and Special Collections, Princeton University Library, Princeton, NJ.

79. "Products of the Future . . . ," *Radio Age* 14, no. 3 (July 1955): 16.

80. "R.C.A. Claims Big TV Advance in Light Device," *Chicago Tribune*, 20 Dec. 1954, C7. Nicoll and Kazan would publish an article summarizing their light-amplification research a year after this initial announcement. See B. Kazan and F. H. Nicoll, "An Electroluminescent Light-Amplifying Picture Panel," *Proceedings of the IRE* 43, no. 12 (Dec. 1955): 1888–97.

81. David Sarnoff, *New Developments in Electronics* (New York: Department of Information, Radio Corporation of America, 1955); Correspondence, David E. Lilienthal Papers, Box 400, 1955 Name Files (O-Sp.); Public Policy Papers, Department of Rare Books and Special Collections, Princeton University Library, Princeton, NJ, 12. See also Robert K. Plumb, "Electronic Device Can Duplicate Every Sound," *New York Times*, 1 Feb. 1955, 35.

82. "R.C.A. Claims Big TV Advance," C7. For more on GE's light amplification project, see Harry C. Kenney, "'Picture on Wall' Envisaged for Television," *Christian Science Monitor*, 23 Dec. 1954, 3.

83. "R.C.A. Claims Big TV Advance," C7.

84. *Dinner in Honor of David Sarnoff (Chairman of the Board, Radio Corporation of America): Commemorating His Fifty Years of Service to Radio, Television and Electronics* (New York: RCA, 1956), 38–57.

85. Ibid., 23.

86. Ibid., 9.

87. David Sarnoff, "Electronic Revolution, Present and Future," *New York Times Magazine*, 30 Sept. 1956, 38.

88. *Dinner*, 27.

89. Ibid., 10–11.

90. Ibid., 24.

91. Ibid.

92. Robert K. Plumb, "R.C.A. Unveils Tape Player, Light Amplifier, Silent Air Cooler," *New York Times*, 2 Oct. 1956, 37.

93. "RCA Unveils Models of Electronic Devices to Honor Sarnoff," *Wall Street Journal*, 2 Oct. 1956, 3.

94. Brown, *And Part of Which I Was*, 269–87.

95. RCA Laboratories, *Research Report 1957*, 27, DSL; B. Kazan, "A Solid-State Amplifying Fluoroscope Screen," *RCA Review* 19, no. 1 (Mar. 1958): 19–34.

96. RCA Laboratories, *Research Report 1956*, p. 2, DSL.

97. Brown, *And Part of Which I Was*, 267–68; Graham, *Business of Research*, 67; Briggs, interview, 12 Oct. 2009.

98. *Dedication*, 25–26.

99. Hyungsub Choi, "The Boundaries of Industrial Research: Making Transistors at RCA, 1948–1960," *Technology and Culture* 48, no. 4 (Oct. 2007): 758–82; Graham, *Business of Research*, 72–73.

100. RCA Laboratories, *Research Report 1955*, 1.

101. RCA Laboratories, *Research Report 1954*, 1; Graham, *Business of Research*, 72.

102. Graham, 71.

103. Kilbon, "Pioneering in Electronics," 243. The RCA color standard, endorsed by the FCC's National Television System Committee (NTSC), established that television pictures would consist of 525 scan lines refreshed at a rate of thirty frames per second. It remained the basis of American color TV broadcasts until 2009, when it was replaced with a new digital transmission standard.

104. Bilby, *The General*, 208.

105. "Faded Rainbow," *Time*, 22 Oct. 1956, 96.

106. Quoted in Bilby, *The General*, 209.

107. Briggs, interview, 12 Oct. 2009.

108. RCA Laboratories, *Research Report 1960*, 54, DSL.

109. Ibid.

110. Bilby, *The General*, 6.

111. The conceit of treating a new technology, or in this instance, the *idea* of a new technology, as a Rorschach test is derived from Sherry Turkle, "'Spinning' Technology: What We Are Not Thinking About When We Are Thinking about Computers,"

in *Technological Visions: The Hopes and Fears That Shape New Technologies*, ed. Marita Sturken et al. (Philadelphia: Temple University Press, 2004), 19–33.

Chapter Two

1. David Sarnoff, "The Fabulous Future," in *The Fabulous Future: America in 1980*, ed. the Editors of *Fortune* (New York: E. P. Dutton & Co., 1956), 19. Sarnoff's article was originally published in the Jan. 1955 issue of *Fortune* but to simplify citation all page numbers will refer to the reprinted essay.

2. Ibid., 21.

3. Ibid., 14.

4. Ibid., 17.

5. Ibid.

6. Alfred R. Zipser, "Personality: Pilot for R.C.A. in Space Field," *New York Times*, 17 May 1959, F3.

7. H. B. Law, "A Review of Some Flat Display Devices," *RCA Engineer* 7, no. 3 (Oct.–Nov. 1961): 20.

8. RCA Laboratories, *Research Report 1957*, 2, 16–18, David Sarnoff Library Collection, Hagley Museum and Library, Wilmington, DE (DSL).

9. Ibid., 3.

10. Ibid., 16.

11. In 1951, W. Ross Aiken, a researcher at the University of California's Radiation Laboratory, reduced the thickness of a CRT by moving its electron gun from its customary position behind the screen to one of its corners and modulating the resulting beam with a series of electrostatic plates. Aiken filed a patent in 1953, and two years later, a DSRC team, including projection TV expert David Epstein (see chap. 1), evaluated his device. See H. S. Allwine, D. C. Darling, C. W. Henderson, S. Lasof, H. B. Law, and E. G. Ramberg, *An Experimental and Theoretical Investigation of a Flat Picture Tube*, PEM-1012C, 4 Oct. 1955, DSL. See also E. G. Ramberg, "Electron-Optical Properties of a Flat Television Picture Tube," *Proceedings of the IRE* 48, no. 12 (Dec. 1960): 1952–60.

12. RCA Laboratories, *Research Report 1955*, 17, DSL; RCA Laboratories, *Research Report 1956*, 10–12, DSL; RCA Laboratories, *Research Report 1957*, 18, DSL.

13. RCA Laboratories, *Research Report 1956*, 26–27, DSL; RCA Laboratories, *Research Report 1957*, 17.

14. J. W. Schwartz, *Line Storage Gas Discharge Picture Reproducer: Practical Approach to Panel Television?*, PEM-719, 28 June 1956, p. 1, DSL. See also George Briggs, interview by author, 12 Oct. 2009, Princeton, NJ; RCA Laboratories, *Research Report 1956*, 27; RCA Laboratories, *Research Report 1957*, 17–18.

15. RCA Laboratories, *Research Report 1957*, 3.

16. Ibid., 17.

17. Jan A. Rajchman, *Mural Television*, PEM-1013C, 26 Oct. 1955, DSL.

18. Kenyon Kilbon, "Pioneering in Electronics: A Short History of the Origins and Growth of RCA Laboratories, Radio Corporation of America, 1919 to 1964" (unpublished typescript, revised Aug. 1964), 166–70, 253–57. Paginated copy available at http://www.davidsarnoff.org/kil.html; Franklin Fisher, James W. McKie, and Richard B. Mancke, *IBM and the U.S. Data Processing Industry: An Economic History* (New York: Praeger, 1983), 71–73; Kenneth Flamm, *Creating the Computer: Government, Industry, and High Technology* (Washington, DC: Brookings Institution, 1988), 123–25.

19. Damon Stetson, "'Brain' Accounts for Army's Gear," *New York Times*, 8 Mar. 1957, 27.

20. Robert Sobel, *RCA* (New York: Stein and Day, 1986), 170–73. See also "Army Buys a Big Brain to Keep Track of Its Tank Parts," *New York Times*, 9 Dec. 1955, 41.

21. A. L. Malcarney quoted in Alfred R. Zipser, "R.C.A. Stepping Up Computer Making," *New York Times*, 14 Apr. 1957, F1. Sales figures from Fisher, McKie, and Mancke, *IBM and the U.S. Data Processing Industry*, 72.

22. "Faded Rainbow," *Time*, 22 Oct. 1956, 96.

23. Kenneth Bilby, *The General: David Sarnoff and the Rise of the Communications Industry* (New York: Harper and Row, 1986), 213. According to Bilby, the $100 million mark was reached in 1957.

24. "RCA Sales Again Top Billion-Dollar Mark," *Radio Age* 16, no. 1 (Jan. 1957): 4. See also "Loss on Color TV to R.C.A. $6,900,000," *New York Times*, 26 Dec. 1956, 38.

25. "Faded Rainbow," 96.

26. "Chasing the Rainbow," *Time*, 30 June 1958, 39.

27. Margaret B. W. Graham, *The Business of Research: RCA and the VideoDisc* (New York: Cambridge University Press, 1986), chaps. 2–3; Sobel, *RCA*, chap. 1.

28. Graham, *Business of Research*, 40–42. See also Alfred D. Chandler Jr., *Inventing the Electronic Century: The Epic Story of the Consumer Electronics and Computer Industries* (New York: Free Press, 2001), 18–21.

29. For more on Zenith's legal battle, which ended with a $10 million settlement in September 1957, see Philip J. Curtis, *The Fall of the U.S. Consumer Electronics Industry* (Westport, CT: Quorum Books, 1994), chaps. 2–3; "Zenith and RCA Settle $61 Million Suit Out of Court," *Wall Street Journal*, 10 Sept 1957, 12. For the Philco suit, see Bilby, *The General*, 214–16; "Philco Suit Cites Electronics 'Plot,'" *New York Times*, 15 Jan. 1957, 37, 40.

30. Anthony Lewis, "RCA Is Indicted in Criminal Case under Trust Law," *New York Times*, 22 Feb. 1958, 1.

31. Consent decrees—legally binding settlements negotiated between the government and the defendant—came into increasing use during the Eisenhower administration as a means of quickly resolving antitrust litigation. For more information on this policy, see Theodore Kovaleff, *Business and Government during the Eisenhower Administration* (Athens: Ohio University Press, 1980), 49–60.

32. "RCA to Open Up 10,000 Radio, TV Patents as It Settles Antitrust Suit," *Wall Street Journal*, 29 Oct. 1958, 3. See also Edward Ranzal, "R.C.A. Yields in Trust Suit; Will Ease Patent Licensing," *New York Times*, 29 Oct. 1958, 1; "Boost for Color TV," *Time*, 10 Nov. 1958, 76.

33. "Boost for Color TV," 76.

34. Bilby, *The General*, 215–16.

35. Jon Gertner, *The Idea Factory: Bell Labs and the Great Age of American Innovation* (New York: Penguin Press, 2012).

36. Ibid., chap. 19.

37. Graham, *Business of Research*, 80–81.

38. Sobel, *RCA*, 26–31, 51–52; "E. J. Nally, 94, Dies; First R.C.A. Head," *New York Times*, 23 Sept. 1953, 31.

39. Bilby, *The General*, 60–62; Sobel, *RCA*, 51–52.

40. Chandler, *Inventing the Electronic Century*, 21–23.

41. Bilby, *The General*, 174–75, 187–89, 213; Sobel, *RCA*, 143–45; Elmer W. Engstrom, "A History of Radio Corporation of America, Part II: The Years 1938–1958," *RCA Engineer* 4, no. 1 (June–July 1958): 28–34.

42. George H. Brown, *And Part of Which I Was: Recollections of a Research Engineer* (Princeton, NJ: Angus Cupar, 1982), 291; Graham, *Business of Research*, 81.

43. "R.C.A. Elects J. L. Burns as President," *Baltimore Sun*, 16 Jan. 1957, 23.

44. Sobel, 144–45.

45. "RCA Elects John Burns, Business Consultant, President and Director," *Wall Street Journal*, 16 Jan. 1957, 8.

46. Bilby, *The General*, 213–14; Graham, *Business of Research*, 81.

47. "RCA's New President," *Radio Age* 16, no. 2 (Apr. 1957): 8.

48. "Boost for Color TV," 76.

49. RCA Laboratories, *Research Report 1957*, 1.

50. Graham, *Business of Research*, 81–83.

51. Zipser, "Personality," F3.

52. RCA Laboratories, *Research Report 1957*, 1.

53. Sobel, *RCA*, 173–75. The 87 percent figure is from Alfred Zipser, "R.C.A. Stepping Up Computer Making," F1.

54. Sobel, *RCA*, 176–77; "R.C.A. and I.B.M. Sign Computer Patents Pact," *New York Times*, 11 Sept. 1957, 45.

55. RCA Electronic Data Processing Division, *RCA 501 Electronic Data Processing System* [1958], 4. PDF available at http://archive.computerhistory.org/resources/text/RCA/RCA.501.1958.102646273.pdf. For more on the RCA 501, see Fisher, McKie, and Mancke, *IBM and the U.S. Data Processing Industry*, 72–73.

56. Alfred R. Zipser, "R.C.A. Sales of Data Systems Running 200% above '59 Level," *New York Times*, 25 Sept. 1960, F1.

57. Elmer W. Engstrom, "A History of Radio Corporation of America, Part III: The Years 1958–1962," *RCA Engineer* 9, no. 1 (June–July 1963): 3.

58. RCA Laboratories, *Research Report 1958*, 1, DSL.

59. RCA Laboratories, *Research Report 1960*, 3, DSL.

60. "Management's Renaissance Man," *Time*, 7 Sept. 1959, 72; Zipser, "Personality," F3.

61. R. K. Lockhart, "Development of a 1000-MC Computer," *RCA Engineer* 6, no. 4 (Dec. 1960–Jan. 1961): 11–13; Jan Rajchman, "RCA Computer Research . . . Some History, and a Review of Current Work," *RCA Engineer* 8, no. 6 (Apr.–May 1963): 8–9; Samuel S. Snyder, "Computer Advances Pioneered by Cryptologic Organizations," *Annals of the History of Computing* 2, no. 1 (Jan. 1980): 60–70. While the US Navy was the nominal sponsor of Project Lightning, Snyder confirms the project was initiated by the National Security Agency with cryptographic applications in mind.

62. RCA Laboratories, *Research Report 1957*, 8; Lockhart, "Development of a 1000-MC Computer," 11.

63. Lockhart, "Development of a 1000-MC Computer," 11–13; Rajchman, "RCA Computer Research," 8–9; Bernard Lechner, interview by author, 9 Apr. 2009, Princeton, NJ.

64. George Briggs, interview by author, 12 Oct. 2009, Princeton, NJ; Lechner, interview, 9 Apr. 2009.

65. RCA Laboratories, *Research Report 1957*, 16–18; RCA Laboratories, *Research Report 1958*, 21–24; RCA Laboratories, *Research Report 1959*, 46–55, DSL; RCA Laboratories, *Research Report 1960*, 49–55; RCA Laboratories, *Research Report 1961*, 48–49, DSL.

66. Law, "Review of Some Flat Display Devices," 23.

67. "How to Install Mural TV," *Popular Science-Special Home Improvement Supplement* 175, no. 3 (Sept. 1959): 174.

68. "Built-in Entertainment," *Electronic Age* 19, no. 2 (Spring 1960): 10.

69. "R.C.A. Shows TVs and Radios of '70s," *New York Times*, 10 May 1961, 73.

70. "RCA Says There Is 'Good Chance' '61 Sales and Net Will Top '60's," *Wall Street Journal*, 2 June 1961, 6.

71. Alfred R. Zipser, "Move by Zenith Hailed by R.C.A.," *New York Times*, 24 Feb. 1961, 37. The $130 million figure comes from Engstrom, "History of Radio Corporation of America, Part III," 2.

72. "RCA Says There Is 'Good Chance,'" 6.

73. "RCA First Period Sales Set Record: Net Trailed '60," *Wall Street Journal*, 3 May 1961, 11.

74. "RCA to Move Headquarters of 2 Units to Indianapolis," *Wall Street Journal*, 9 Aug. 1960, 4. For more on BMEWS and TIROS, see Engstrom, "History of Radio Corporation of America, Part III," 3–4.

75. Graham, *Business of Research*, 83.

76. Brown, *And Part of Which I Was*, 291.

77. Fisher, McKie, and Mancke, *IBM and the U.S. Data Processing Industry*, 74. The executive quoted here is Edwin McCollister, a vice president in IEP's Electronic Data Processing division, as confirmed in Sobel, *RCA*, 178.

78. See, e.g., John L. Burns, "Forty Years of Electronic Progress," *Electronic Age* 18, no. 3 (Summer 1959): 2–3; "The Three C's . . . Electronics' Newest Surge," *Electronic Age* 19, no. 2 (Spring 1960): 13.

79. Brown, *And Part of Which I Was*, 297.

80. "Burns Quits R.C.A. Presidency; Elmer W. Engstrom Is Elected," *New York Times*, 2 Dec. 1961, 28; "Engstrom's Selection as RCA's President Termed Move to Boost Lagging Earnings," *Wall Street Journal*, 4 Dec. 1961, 8.

81. "Engstrom's Selection as RCA's President," 8.

82. "Dr. Elmer W. Engstrom Elected President of RCA," *RCA Engineer* 7, no. 4 (Dec. 1961–Jan. 1962): center spread (between pp. 30 and 31]

83. Fisher, McKie, and Mancke, *IBM and the U.S. Data Processing Industry*, 73–75, 202.

84. RCA Laboratories, *Research Report 1962*, 1, DSL.

85. Radio Corporation of America, *Working in Research at RCA* (Princeton, NJ: RCA Laboratories Division, 1956), 9; RCA Laboratories, *Research Report 1960*, 6.

86. Graham, *Business of Research*, 84–85; Kilbon, "Pioneering in Electronics," 340–41.

87. RCA Laboratories, *Research Report 1961*, 7, DSL.

88. RCA Laboratories, *Research Report 1962*, 9.

89. Ibid.

90. Tom Howe, "Jay Brandinger," *Memories of VideoDisc: Who's Who in VideoDisc*, CEDmagic.com, http://www.cedmagic.com/mem/whos-who/brandinger-jay.html. RCA's Long Island laboratories were the remnants of the company's earliest radio research facilities, founded in the 1920s. For more information, see Kilbon, "Pioneering in Electronics," chap. 1; J. C. Warner, "A History of Radio Corporation of America: The Years to 1938," *RCA Engineer* 3, no. 1 (June–July 1957): 3–9.

91. Brown, *And Part of Which I Was*, 120, 132–35, 151; Kilbon, "Pioneering in Electronics," 214–29; John van Raalte, interview by author, 3 June 2009, Princeton, NJ.

92. RCA Laboratories, *Research Report 1962*, 9.

93. Van Raalte, interview, 3 June 2009. This characterization is further confirmed in William Webster, telephone conversation with author, 21 Apr. 2010; Graham, *Business of Research*, 70–71.

94. RCA Laboratories, *Research Report 1963*, 26, DSL. For more details on the origins and operation of the Eidophor, see Larry J. Hornbeck, "From Cathode Rays to Digital Micromirrors: A History of Electronic Projection Display Technology," *TI Technical Journal* (July–Sept. 1998): 11–12.

95. John van Raalte, *SLV: A New TV Projection System*, PTR-2698, 27 Mar. 1969, DSL, 1. SLV stood for "Schlieren light valve," a reference to the type of optical system used to transform the distortion patterns into a television image.

96. RCA Laboratories, *Research Report 1963*, 26; RCA Laboratories, *Research Report 1964*, 30–31, DSL; RCA Laboratories, *Research Report 1965*, 30–31, DSL; RCA Laboratories, *Research Report 1966*, 24, DSL.

97. D. H. Pritchard, PTR-2205: *Electro-Optic Light Valve TV Display*, 4 Jan. 1967, DSL.

98. RCA Laboratories, *Research Report 1966*, 9.

99. Van Raalte, interview, 3 June 2009.

100. RCA Laboratories, *Research Report 1963*, 27; Bruce Shore, "Lasers: New Power From Light," *Electronic Age* 21, no. 3 (Summer 1962): 8–11.

101. RCA Laboratories, *Research Report 1966*, 10, 25–26. See also RCA Laboratories, *Research Report 1965*, 32; Van Raalte, interview, 3 June 2009.

102. RCA Laboratories, *Research Report 1966*, 9. See, e.g., the "flat-panel vacuum display" discussed on pp. 24–25 of this report, which used a matrix array of integrated circuits to address a series of electron-excited phosphor elements.

103. RCA Laboratories, *Research Report 1966*, 9.

104. John van Raalte would assume responsibility for refining Kell's deformable film system. See Van Raalte, *SLV*. The introduction of the DSRC's 1966 research report indicates Pritchard's display was transferred to RCA's Defense Electronics Product division in Camden at the end of that year.

105. Fisher, McKie, and Mancke, *IBM and the U.S. Data Processing Industry*, 74–75; RCA Laboratories, *Research Report 1962*, 39.

106. RCA Laboratories, *Research Report 1962*, 23–24, 39–40.

107. Lechner, interview, 9 Apr. 2009. For more on Lechner's background and early days at RCA, see Bernard Lechner, interview by Alexander Magoun, 25 May 2004, Princeton, NJ.

108. Bernard Lechner, Notebook 3775, 4084, and 8593, DSL.

109. Law, "Review of Some Flat Display Devices," 20–23; Jess Josephs, "A Review of Panel-Type Display Devices," *Proceedings of the IRE* 48, no. 8 (Aug. 1960): 1380–95; E. A. Sack, "ELF: A New Electroluminescent Display," *Proceedings of the IRE* 46, no. 10 (Oct. 1958): 1694–99.

110. RCA's Zurich lab opened in 1955 to encourage closer contact with European scientists and patent licensees. For further information, see C. G. Mayer, "Laboratories RCA, Ltd.," *Radio Age* 15, no. 1 (Jan. 1956): 20–21.

111. Lechner, interview, 9 Apr. 2009; Ennio Fatuzzo, Notebook 15788, DSL; RCA Laboratories, *Research Report 1962*, 157.

112. Rajchman, *Mural Television*, 11; RCA Laboratories, *Research Report 1954*, 72, DSL. See also Jan A. Rajchman and George R. Briggs, Electroluminescent Apparatus, US Patent 3,041,490, filed 31 May 1955 and issued 26 June 1962.

113. C. F. Pulvari, "The Transpolarizer: An Electrostatically Controlled Circuit Impedance with Stored Setting," *Proceedings of the IRE* 47, no. 6 (June 1959): 1123.

114. Lechner, interview, 9 Apr. 2009.

115. B. J. Lechner et al., *Development of a Solid-State Matrix Display*, AFFDL-TR-67-71, July 1967, ii, Defense Technical Information Center, Ft. Belvoir, VA. See also Lechner, interview, 9 Apr. 2009; B. J. Lechner et al., *Preliminary Development of a Solid State Matrix Display*, AFFDL-TR-66-5, Jan. 1967, Defense Technical Information Center, Ft. Belvoir, VA.

116. Lechner, interview, 25 May 2004.

117. George Taylor, Notebook 18717, p. 44, DSL.

118. George Taylor, interview by author, 14 Oct. 2009, Pennington, NJ; RCA Laboratories, *Research Report 1964*, 121–22.

119. Lechner et al., *Preliminary Development*, 60–83; RCA Laboratories, *Research Report 1965*, p. 57–58, DSL. Lechner used the term *exerciser* to refer to these displays in his 9 Apr. 2009 interview with the author.

120. Lechner, interview, 9 Apr. 2009.

121. Juri Tults, Notebook 29309, p. 46, DSL.

122. Full technical details can be found in Lechner et al., *Development*, 64–97 as well as George W. Taylor, "The Design and Operating Characteristics of a 1200-Element Ferroelectric-Electroluminescent Display," *IEEE Transactions on Electron Devices* ED-16, no. 6 (June 1969): 565–75.

123. George Taylor, Notebook 29288, p. 88, DSL.

124. Tults, Notebook 29309, p. 47; Lechner, interview, 25 May 2004; Taylor, interview, 14 Oct. 2009.

125. Lechner et al., *Development*, 98.

126. RCA Laboratories, *Research Report 1966*, 7.

127. Sarnoff, "Fabulous Future," 20.

128. Lechner, interview, 9 Apr. 2009.

Chapter Three

1. For more on this research, see L. A. Barton and W. H. Moles, *Memory Characteristics of Photoconductor Materials*, PEM-2484, 6 Aug. 1964; RCA Laboratories, *Research Report 1963*, 24–25; RCA Laboratories, *Research Report 1964*, 29, all in David Sarnoff Library Collection, Hagley Museum and Library, Wilmington, DE (DSL).

2. Lucian Barton, Notebook 24188, p. 85, DSL.

3. Ibid.

4. For more on Zworykin's career, see Albert Abramson, *Zworykin, Pioneer of Television* (Urbana: University of Illinois Press, 1995).

5. George H. Heilmeier, "Liquid Crystal Displays: An Experiment in Interdisciplinary Research That Worked," *IEEE Transactions on Electron Devices* ED-23, no. 7 (July 1976): 782 (emphasis in original).

6. E.g., Heilmeier, "Liquid Crystal Displays," 782; Louis Zanoni, interview by author, 1 July 2009, Ewing, NJ.

7. Richard Williams, Electro-Optical Elements Utilizing an Organic Nematic Compound, US Patent 3,322,485, filed 9 Nov. 1962 and issued 30 May 1967.

8. George Heilmeier, interview by Margaret Dennis and Carlene Stephens, 4 Dec. 1998, Morristown, NJ. See also Amy Calhoun, "George Heilmeier," *Penn Engineering* (Fall 2009): 10–11.

9. George H. Heilmeier, "An Analysis of Parametric Amplification in Periodically Loaded Transmission Lines" (MS thesis, Princeton University, 1960).

10. Heilmeier, "Liquid Crystal Displays," 780.

11. George Heilmeier, Notebook 14445, front index, DSL.

12. Heilmeier, "Liquid Crystal Displays," 780.

13. For a general history of organic electronics, see Tarek Zaki, *Short-Channel Organic Thin-Film Transistors* (Cham, Switzerland: Springer, 2015), 5–11. For more on the development of OLEDs, see Stephen Forrest, Paul Burrows, and Mark Thompson, "The Dawn of Organic Electronics," *IEEE Spectrum* 37, no. 8 (Aug. 2000): 29–34; Mitsuhiro Koden, *OLED Displays and Lighting* (Hoboken, NJ: John Wiley and Sons, 2016), 1–11. For more on organic solar cells, see Peter Fairley, "Solar-Cell Squabble," *IEEE Spectrum* 45, no. 4 (Apr. 2008): 36–40.

14. Heilmeier, interview, 4 Dec. 1998. For more on Harrison's early phthalocyanine research, see RCA Laboratories, *Research Report 1960*, 122–23, DSL.

15. George H. Heilmeier, "Semiconduction in Phthalocyanine: A Study in Organic Semiconduction" (PhD diss., Princeton University, 1962), vii.

16. Christophe Lécuyer and David C. Brock, *Makers of the Microchip: A Documentary History of Fairchild Semiconductor* (Cambridge, MA: MIT Press, 2010), 150–56.

17. Harry Kihn, "Integrated Electronics . . . A Survey," *RCA Engineer* 7, no. 6 (Apr.–May 1962): 14–19. For more on Weimer's thin-film transistor research, see P. K. Weimer, "The TFT . . . A New Thin-Film Transistor," *RCA Engineer* 7, no. 6 (Apr.–May 1962): 20–22.

18. S. E. Harrison et al., *Utilization of Organic Materials for Molecular Electronics: Technical Documentary Report no. AL TDR 64-178*, July 1964, p. iii, DSL. For more on molecular electronics, see Hyungsub Choi and Cyrus C. M. Mody, "The Long History of Molecular Electronics: Microelectronics Origins of Nanotechnology," *Social Studies of Science* 39, no. 1 (Feb. 2009): 11–50.

19. Harrison et al., *Technical Documentary Report*, iii.

20. George Heilmeier, Notebook 20858, DSL.

21. Heilmeier, "Liquid Crystal Displays," 780.

22. RCA Laboratories, *Research Report 1960*, 66–67.

23. D. J. Blattner and F. Sterzer, "Modulators and Demodulators for Laser Systems," *RCA Engineer* 8, no. 5 (Feb.–Mar. 1963): 16–19.

24. George Heilmeier, Interview with Margaret Dennis and Carlene Stephens, Morristown, NJ, 4 Dec. 1998.

25. George H. Heilmeier and Wolfgang Liebmann, PEM-2297: *A Note on the Crystal*

Growth and Evaluation of Hexamine as an Optical Modulator, 9 May 1963, DSL, 1–3. Liebmann, a semiconductor researcher, grew the crystals for Heilmeier's experiments before leaving RCA at the end of 1963.

26. Heilmeier, Notebook 20858, pp. 47–66; George Heilmeier, Notebook 22280, pp. 1–42, DSL; Harrison et al., *Technical Documentary Report*, 11–17, 28–35.

27. Heilmeier, "Liquid Crystal Displays," 780–81.

28. Heilmeier, Notebook 22280, p. 7. The terms *guest* and *host* are first used to refer to solutes and solvents with reference to this application on p. 66 of the same notebook.

29. Heilmeier, Notebook 22280, pp. 11–13, 41–42, 57–59, 66–68, 82–84.

30. George Heilmeier, Notebook 24160, p. 72, DSL.

31. Ibid.

32. Ibid., 100.

33. Ibid.

34. Richard Williams, Interview by author, 8 Apr. 2009, Princeton, NJ.

35. Ibid.

36. Williams, interview, 8 Apr. 2009; Richard Williams and Richard H. Bube, "Photoemission in the Photovoltaic Effect in Cadmium Sulfide Crystals," *Journal of Applied Physics* 31, no. 6 (June 1960): 968–78.

37. Richard Williams, interview by author, 16 Apr. 2009, Princeton, NJ; "Electric Field Induced Light Absorption in CdS," *Physical Review* 117, no. 6 (15 Mar. 1960): 1487–90; "Effect of a High Electric Field on the Absorption of Light by PbI_2 and HgI_2," *Physical Review* 126, no. 2 (15 Apr. 1962): 442–46.

38. Williams, interview, 16 Apr. 2009.

39. Ibid.

40. The National Television System Committee (NTSC) established that in the United States, television images would consist of 525 scanning lines. To reduce flicker, the odd- and even-numbered lines were transmitted in two consecutive fields and interlaced with one another to produce a complete picture. Since 30 frames per second were needed to ensure the illusion of motion, the fields had to be transmitted at a frequency of 60 Hz, much lower than the 5 kHz switching speeds reported in Williams, "Effect of a High Electric Field," 445.

41. Williams, interview, 16 Apr. 2009.

42. Ibid.

43. For more on the early history of liquid crystal research, see David Dunmur and Tim Sluckin, *Soap, Science, and Flat-Screen TVs: A History of Liquid Crystals* (New York: Oxford University Press, 2011; repr., 2014), chaps. 2–3. Reinitzer and Lehmann's respective articles on liquid crystals and many other primary sources on the subject are collected in Timothy J. Sluckin, David A. Dunmur, and Horst Stegemeyer, eds., *Crystals That Flow: Classic Papers from the History of Liquid Crystals* (New York: Taylor and Francis, 2004).

44. Dunmur and Sluckin, *Soap, Science, and Flat-Screen TVs*, chap. 4, esp. 80–90; Georges Friedel, Les états mésomorphes de la matière," *Annales des Physique* 18 (1922): 273–474, trans. "The Mesomorphic States of Matter," in Sluckin et al., *Crystals That Flow*, 162–211. Friedel's article (Sluckin et al., *Crystals That Flow*, 167) is also the source of the estimated number of liquid crystals in 1908 referenced in the previous paragraph.

45. Physicists use the term *isotropic* to describe a substance whose properties remain consistent regardless of physical orientation. *Anisotropic* materials, including the liquid crystals discussed here, have properties that are direction-dependent.

46. James L. Fergason, "Liquid Crystals," *Scientific American* 211, no. 2 (Aug. 1964): 77. Although George Gray published the first English-language volume exclusively dedicated to liquid crystals in 1962, the DSRC library only received a copy that August, several months after Williams's experiments. See G. W. Gray, *Molecular Structure and the Properties of Liquid Crystals* (New York: Academic Press, 1962); David Sarnoff Library, *Accession Records*, 24 Aug. 1962, DSL.

47. Fergason, "Liquid Crystals," 85; Mary Knoblauch, "Mood Ring Monitors Your State of Mind," *Chicago Tribune*, 8 Oct. 1975, C1. For more on Fergason's liquid crystal research, see Terri Fergason Neal and Marian Pierce, *Mr. Liquid Crystal: The Untold Story of How James L. Fergason Invented the Liquid Crystal Display and Helped Create the Digital World* (Los Angeles: New Insights Press, 2016).

48. Williams, interview, 16 Apr. 2009.

49. Richard Williams, Notebook 15811, p. 75, DSL.

50. Ibid.

51. Ibid., 77.

52. Williams, interview, 16 Apr. 2009.

53. Richard Williams, interview by author, 30 Apr. 2009, Princeton, NJ.

54. Williams, Notebook 15811, pp. 78, 87–95; Richard Williams, Notebook 17672, pp. 1–14, DSL.

55. Williams, US Patent 3,322,485, p. 1.

56. Barnett Levin and Nyman Levin, "Improvements in or Relating to Light Valves," UK Patent 441,724, filed 13 July 1934 and issued 13 Jan. 1936.

57. Williams, US Patent 3,322,485, p. 8.

58. Williams, interview, 16 Apr. 2009.

59. Richard Williams, Notebook 20170, pp. 1–56, DSL.

60. Simon Larach, Notebook 982, pp. 65 (original punctuation), DSL.

61. Williams, interview, 16 Apr. 2009. William Webster noted in a 3 Feb. 2011 telephone conversation with the author that RCA's patent attorneys were also familiar with x-y addressing and might have discussed the concept with Williams as they drafted the original disclosure form.

62. Bernard Lechner, interview by author, 17 Apr. 2009, Princeton, NJ.

63. Richard Williams, "High Electric Fields in Sodium Chloride," *Journal of Physics and Chemistry of Solids* 25, no. 8 (Aug. 1964): 853–58. At the end of this article, Williams notes that his work was sponsored by the US Army Research Office.

64. Richard Williams, "Domains in Liquid Crystals," *Journal of Chemical Physics* 39, no. 2 (15 July 1963): 384–88; "Liquid Crystals in an Electric Field," *Nature* 199 (20 July 1963): 273–74.

65. Williams, interview, 30 Apr. 2009.

66. Heilmeier, interview, 4 Dec. 1998.

67. George H. Heilmeier, "Transient Behavior of Domains in Liquid Crystals," *Journal of Chemical Physics* 44, no. 2 (15 Jan. 1966): 647; G. H. Heilmeier, L. A. Zanoni, and L. A. Barton, "Dynamic Scattering in Nematic Liquid Crystals," *Applied Physics Letters* 13, no. 1 (1 July 1968): 47.

68. Heilmeier, Notebook 24160, p. 100, DSL. See also Richard Williams and George Heilmeier, "Possible Ferroelectric Effects in Liquid Crystals and Related Liquids," *Journal of Chemical Physics* 44, no. 2 (15 Jan. 1966): 638–43.

69. Heilmeier, interview, 4 Dec. 1998.

70. Zanoni, interview, 1 July 2009.

71. G. H. Heilmeier, "The Dielectric and Electrooptical Properties of a Molecular Crystal: Hexamine." *Applied Optics* 3, no. 11 (Nov. 1964): 1287.

72. George Heilmeier, Notebook 25463, pp. 39–40, DSL.

73. Ibid., 44, 56.

74. Zanoni, interview, 1 July 2009.

75. Joseph Castellano, *Liquid Gold: The Story of Liquid Crystal Displays and the Creation of an Industry* (Hackensack, NJ: World Scientific, 2005), 24.

76. Heilmeier, Notebook 25463, pp.58–59; Joel Goldmacher, Notebook 25011, p. 14, DSL.

77. Heilmeier, Notebook 25463, pp. 57–60 (quotation from p. 60).

78. Heilmeier, Notebook 26848, 2–6; Warren Moles, Notebook 26365, p. 10. See also G. H. Heilmeier and L. A. Zanoni, "Guest-Host Interactions in Nematic Liquid Crystals: A New Electro-Optic Effect," *Applied Physics Letters* 13, no. 3 (1 Aug. 1968): 91–92.

79. Heilmeier, Notebook 25463, pp. 61–62.

80. Louis Zanoni, Notebook 16971, p. 93, DSL.

81. Heilmeier, "Liquid Crystal Displays," 781.

82. Ibid.

83. RCA Laboratories, *Research Report 1964*, 4.

84. Lechner, interview, 17 Apr. 2009; Naomi Kleitman, e-mail message to author, 9 May 2014.

85. John van Raalte, interview by author, 3 June 2009, Princeton, NJ.

86. Barton, Notebook 24188, pp. 84–100; Moles, Notebook 26365, DSL, pp. 7–64.

87. Moles, Notebook 26365, p. 41.

88. Heilmeier, "Liquid Crystal Displays," 782.

89. Heilmeier, Notebook 26848, pp. 1–60, DSL.

90. Ibid., 44. DHOBABB's full name was 3-3′-dichloro-4,4′-di (p-n-hepytloxybenzylidene-amino) biphenyl.

91. Heilmeier, Notebook 26848, pp. 44, 47, DSL.

92. Heilmeier, Notebook 25463, p. 96; Notebook 26848, pp. 10–11, 18; Williams, "Domains in Liquid Crystals," 385. In his article, Williams also noted that the "stirring action of an electric field on liquid crystals" had been observed by researchers in the Soviet Union as early as 1937.

93. Heilmeier, Notebook 26848, pp. 44–45.

94. Ibid., 47.

95. Ibid.

96. Ibid., 97.

97. Ibid. BBA was an abbreviation for p-n butoxybenzoic acid.

98. George Heilmeier, Notebook 28088, p. 3, DSL.

99. Ibid, 12.

100. Lechner, interview, 17 Apr. 2009.

101. Van Raalte, interview, 3 June 2009.

102. RCA Laboratories, *Progress Report: Liquid Crystals, Mar.–May 1965*, 6, DSL.

103. R. E. Quinn, "Interdepartmental Correspondence: Liquid Crystal Display Report—January–March 1967," 1 June 1967, DSL, 1. In interviews by the author, Bernard Lechner and John van Raalte (17 Apr. 2009 and 3 June 2009, respectively) referenced two other projects that earned company-private status—the penetration phosphor CRT and the line-screen beam index tube—both of which could have radically altered color TV production. Given RCA's substantial investment in CRT televisions, it is likely that the DSRC's management viewed these efforts, at least initially, as defensive in nature. David Kleitman appears to have been involved in both of these initiatives, which may explain how he was able to secure "company-private" status for the LCD. Naomi Kleitman, e-mail message to author, 12 May 2014.

104. Castellano, *Liquid Gold*, 21–22; Lechner, interview, 17 Apr. 2009; Williams, interview, 30 Apr. 2009.

105. "International Liquid Crystal Conference," program, Kent State University, 1965, Liquid Crystal Institute Records, 1965–1998, Box 1, Folder 6, Kent State University Special Collections. Two other members of the RCA liquid crystal team, Joel Goldmacher and Joseph Castellano, also attended this conference, but neither delivered a lecture. Castellano, *Liquid Gold*, 34–39.

106. As historian of engineering Ann Johnson points out, this type of negotiation over the lines between proprietary and public knowledge is typical within corporate settings. See Ann Johnson, *Hitting the Brakes: Engineering Design and the Production of Knowledge* (Durham, NC: Duke University Press, 2009), 138–39. For examples of

fundamental liquid crystal studies written before RCA decided to go public with the LCD, see Joel Goldmacher and Lucian A. Barton, "Liquid Crystals. I. Fluorinated Anils," *Journal of Organic Chemistry* 32, no. 2 (Feb. 1967): 476–77; Heilmeier, "Transient Behavior."

107. Van Raalte, interview, 3 June 2009; Quinn, "Interdepartmental Correspondence," 1.

108. Van Raalte, interview, 3 June 2009; Naomi Kleitman, e-mail message to author, 12 May 2014. Further information concerning Kleitman's departure is based on internal organizational charts and his obituary: "Dr. David Kleitman," *San Francisco Chronicle* (25 Feb. 2004), http://www.sfgate.com/news/article/KLEITMAN-Dr-David-2818193.php. For more on Signetics's contributions to the semiconductor industry, see Christophe Lécuyer, *Making Silicon Valley: Innovation and the Growth of High Tech, 1930–1970* (Cambridge, MA: MIT Press, 2006), chap. 6.

109. Heilmeier, interview, 4 Dec. 1998; Louis Zanoni, interview by author, 21 Aug. 2009, Ewing, NJ.

110. Barton, Notebook 24188, pp. 84–100; RCA Laboratories, *Progress Report: Liquid Crystals, Jun.–Jul. 1965*, 1–4, DSL; Castellano, *Liquid Gold*, 23–24.

111. Lécuyer and Brock, *Makers of the Microchip*.

112. Castellano, *Liquid Gold*, 24–25.

113. RCA Laboratories, *Progress Report: Liquid Crystals, Mar.–May 1965*, 6; Jean Kane, e-mail message to author, 16 July 2009.

114. For more on Cold War concern about the relationship between science, the military, and industry, see Kelly Moore, *Disrupting Science: Social Movements, American Scientists, and the Politics of the Military, 1945–1975* (Princeton, NJ: Princeton University Press, 2008); Matthew Wisnioski, *Engineers for Change: Competing Visions of Technology in 1960s America* (Cambridge, MA: MIT Press, 2012).

115. Castellano, *Liquid Gold*, 11.

116. RCA Laboratories, *Progress Report: Liquid Crystals, Jun.–Jul. 1965*, 1, DSL.

117. Ibid., 1–2; Joel Goldmacher, Notebook 27361, DSL; Castellano, *Liquid Gold*, 25.

118. RCA Laboratories, *Progress Report: Liquid Crystals, Jan.–Mar. 1966*, 19, DSL. The "12" in "APAPA-12" referred to the total number of carbon atoms in the modified APAPA's alkoxy group (1) and the number of carbon atoms added to its acyloxy group (2). See RCA Laboratories, *Progress Report: Liquid Crystals, Jan.–Mar. 1967*, 2, DSL.

119. RCA Laboratories, *Progress Report: Liquid Crystals, Oct. 1965*, 3–4, DSL. The liquid crystal mixture mentioned here was a fifty-fifty mix of APAPA-10 (nematic range 82°C–110°C) and APAPA-40 (nematic range 82°C–113°C) with a new nematic range of 47°C–108°C.

120. RCA Laboratories, *Progress Report: Liquid Crystals, Oct. 1965*, 6, DSL.

121. RCA Laboratories, *Progress Report: Liquid Crystals, Apr. 1966*, 2; Joseph Castellano, Notebook 28304, p. 76, both in DSL.

122. RCA Laboratories, *Progress Report: Liquid Crystals, Jan.–Mar. 1967*, 3–6; *Progress Report: Liquid Crystals, Apr.–Jun. 1967*, 2–4, 8–11, both in DSL.

123. RCA Laboratories, *Progress Report: Liquid Crystal Progress Report, Jan.–Mar. 1966*, 4–6. See also Jean Kane, Notebook 28090, pp. 59–100; Notebook 31729, both in DSL.

124. Heilmeier, Notebook 28088, p. 3; RCA Laboratories, *Progress Report: Liquid Crystals, Mar.–May 1965*, 4–5, DSL.

125. RCA Laboratories, *Progress Report: Liquid Crystals, Dec. 1965*, 5, DSL (emphasis in original).

126. George Heilmeier, Notebook 30812, p. 1, DSL.

127. RCA Laboratories, *Progress Report: Liquid Crystals, Jun. 1966*, 4, DSL (emphasis in original).

128. Heilmeier, Notebook 30812, p. 58.

129. Heilmeier, Notebook 28088, 9.

130. RCA Laboratories, *Progress Report: Liquid Crystals, Sep. 1965*, 5, DSL.

131. RCA Laboratories, *Progress Report: Liquid Crystals, Jun. 1966*, 5–6, DSL. See also George H. Heilmeier, Louis A. Zanoni, and Lucian A. Barton, "Dynamic Scattering: A New Electrooptic Effect in Certain Classes of Nematic Liquid Crystals," *Proceedings of the IEEE* 56, no. 7 (July 1968): 1162–71.

132. RCA Laboratories, *Progress Report: Liquid Crystals, Aug. 1966*, 5–6, DSL.

133. RCA Laboratories, *Progress Report: Liquid Crystals, Jan.–Mar. 1967*, 9–10, DSL.

134. RCA Laboratories, *Progress Report: Liquid Crystals, Apr.–Jun. 1967*, 13–15, DSL.

135. George Heilmeier, interview by Stewart Finley, *Men and Molecules*, episode 515, "Liquid Crystals: A Bright Promise," ca. 1970.

136. In the more formal parlance of physical chemistry, APAPA exhibited negative dielectric anisotropy. Nematic compounds that rotated with their long axes parallel to the field possessed positive dielectric anisotropy and would not produce dynamic scattering.

137. RCA Laboratories, *Progress Report: Liquid Crystals, Dec. 1966*, 6–8, DSL.

138. Alan Sussman, interview by author, 19 Jan. 2011, Princeton, NJ. For more on the work listed here, see RCA Laboratories, *Progress Report: Liquid Crystals, Oct–Dec. 1967*, 17; *Progress Report: Liquid Crystals, Jan.–Mar. 1968*, 8–11, both in DSL.

139. Wolfgang Helfrich, interview by author, 28 Sept. 2009, Berlin, Germany; RCA Laboratories, *Progress Report: Liquid Crystals, Oct.–Dec. 1967*, 10, DSL.

140. Heilmeier, Notebook 26848, pp. 24–28.

141. RCA Laboratories, *Progress Report: Liquid Crystals, Aug. 1965*, 5, DSL. The 250,000 display element figure is derived from RCA Laboratories, *Progress Report: Liquid Crystals, Jun. 1966*, 1.

142. RCA Laboratories, *Progress Report: Liquid Crystals, Nov. 1965*, 6–7, DSL. For more on the photolithographic techniques outlined above, see Lécuyer and Brock, *Makers of the Microchip*, 19–20.

143. RCA Laboratories, *Progress Report: Liquid Crystals, Jan.–Mar. 1966*, 8–9, 15–16; Moles, Notebook 26365, pp. 97–100; John van Raalte, Notebook 27689, pp. 98–100; John van Raalte, Notebook 30300, pp. 1–3, all in DSL.

144. Heilmeier, Notebook 30812, pp. 32–33; Louis Zanoni, Notebook 27707, pp. 60–61, DSL.

145. RCA Laboratories, *Progress Report: Liquid Crystals, Apr. 1966*, 9, DSL. See also Heilmeier, Notebook 30812, pp. 28–31, and Zanoni, Notebook 27707, pp. 61–62, both in DSL.

146. RCA Laboratories, *Progress Report: Liquid Crystals, Jun. 1966*, 2, DSL.

147. RCA Laboratories, *Progress Report: Liquid Crystals, Nov. 1966*, 4, DSL; John van Raalte, Notebook 32465, pp. 18–29, DSL. Technical details of this system can be found in: John A. van Raalte, "Reflective Liquid Crystal Television Display," *Proceedings of the IEEE* 56, no. 12 (Dec. 1968): 2146–49.

148. Van Raalte, Notebook 32465, p. 29.

149. Van Raalte, interview, 3 June 2009.

150. RCA Laboratories, *Progress Report: Liquid Crystals, Nov. 1966*, 9, DSL.

151. Lechner, interview, 17 Apr. 2009. Lechner also described his LCD research in "History Crystallized: A First-Person Account of the Development of Matrix-Addressed LCDs for television at RCA in the 1960s," *Information Display* (Jan. 2008): 26–30.

152. RCA Laboratories, *Progress Report: Liquid Crystals, Nov. 1966*, 9, DSL; George Heilmeier, Notebook 32284, p. 51, DSL; Lechner, interview, 17 Apr. 2009.

153. RCA Laboratories, *Progress Report: Liquid Crystals, Jan.–Mar. 1967*, 19–22; *Progress Report: Liquid Crystals, Apr.–Jun. 1967*, 30, both in DSL; Lechner, "History Crystallized," 28–29.

154. RCA Laboratories, *Progress Report: Liquid Crystals, Apr.–Jun. 1967*, 31–32, DSL.

155. RCA Laboratories, *Progress Report: Liquid Crystals, Oct.–Dec. 1967*, 26–30, DSL; Lechner, "History Crystallized," 29.

156. Lechner and his team summarized their LCD research in Bernard J. Lechner, Frank J. Marlowe, Edward O. Nester, and Juri Tults, "Liquid Crystal Matrix Displays," *Proceedings of the IEEE* 59, no. 11 (Nov. 1971): 1566–79.

157. RCA Laboratories, *Progress Report: Liquid Crystals, Jan.–Mar. 1968*, 27, DSL.

158. Bernard Lechner, interview by author, 24 Apr. 2009, Princeton, NJ; Lechner, "History Crystallized," 29–30.

159. RCA Laboratories, *Progress Report: Liquid Crystals, Jan.–Mar. 1967*, 11–13; *Progress Report: Liquid Crystals, Jan.–Mar. 1968*, 19–20; A. Pinsky, "Preliminary Plans for Press Conference to Announce RCA Development of Liquid Crystal Electron Optic [*sic*] Effect," interdepartmental correspondence, 10 Apr. 1968, Bernard Lechner Collection, all in DSL.

160. George H. Heilmeier, Louis A. Zanoni, and Lucian A. Barton, *Dynamic*

Scattering . . . A New Electro-Optic Effect in Certain Classes of Nematic Liquid Crystals, PTR-2366, 1 Nov. 1967, DSL; Heilmeier, Zanoni, and Barton, "Dynamic Scattering."

161. RCA Laboratories, *Research Report 1967*, 7–8, 93, DSL. Heilmeier would later compose a more detailed summary for *RCA Engineer*; see G. H. Heilmeier, "Liquid Crystals: The First Electronic Method for Controlling the Reflection of Light," *RCA Engineer* 15, no. 1 (June–July 1969): 14–18.

162. RCA Laboratories, *Progress Report: Liquid Crystals, Oct.–Dec. 1967*, DSL.

163. J. F. Spivak to J. V. Regan, "Patent Coverage Related to Liquid Crystals," 27 June 1968, p. 3, DSL.

164. Richard Williams, Notebook 34286; RCA Laboratories, *Research Report 1967*, 133–38; RCA Laboratories, *Research Report 1968*, 62–63, all in DSL.

165. Williams, interview, 16 Apr. 2009.

166. Ibid.

167. Nicholas King, "Young Scientists of Today," *Electronic Age* 28, no. 3 (Summer 1969): 30.

168. Isaiah Berlin, *The Hedgehog and the Fox: An Essay on Tolstoy's View of History* (London: Weidenfeld and Nicolson, 1953), 1.

169. Thomas P. Hughes, *Networks of Power: Electrification in Western Society, 1880–1930* (Baltimore: Johns Hopkins University Press, 1983), 18.

170. For more on the importance of individual research styles in the history of science and technology, see Thomas P. Hughes, *American Genesis: A Century of Technological Enthusiasm, 1870–1970* (1989; repr., Chicago: University of Chicago Press, 2004); W. Bernard Carlson, *Tesla: Inventor of the Electrical Age* (Princeton, NJ: Princeton University Press, 2013); Alan J. Rocke, *Image and Reality: Kekulé, Kopp, and the Scientific Imagination* (Chicago: University of Chicago Press, 2010).

171. For more on innovation "champions," see Jane M. Howell and Christopher A. Higgins, "Champions of Change: Identifying, Understanding, and Supporting Champions of Technological Innovations," *Organizational Dynamics* 19, no. 1 (Summer 1990): 40–55; Morell E. Mullins et al., "The Role of the Idea Champion in Innovation: The Case of the Internet in the Mid-1990s," *Computers in Human Behavior* 24, no. 2 (Mar. 2008): 451–67.

Chapter Four

1. RCA Corporation, *Annual Report 1968* (New York: RCA, 1969), 3–5.

2. Robert Sobel, *RCA* (New York: Stein and Day, 1986), 184–86.

3. Lynn Taylor, "R.C.A. Extinguishes Trademark Flash," *Chicago Tribune*, 19 Jan. 1968, C8. RCA retained the new trademark until its 1986 sale to General Electric. GE later transferred RCA's television business and associated branding to Thomson SA, a French firm that proceeded to license the RCA logo to a variety of businesses.

4. Sobel, *RCA*, 187–91; Kenneth Bilby, *The General: David Sarnoff and the Rise of the Communications Industry* (New York: Harper and Row, 1986), 276–78.

5. RCA Corporation, *Annual Report 1968*, 4.

6. Ibid., 3.

7. Ibid., 5.

8. Ibid.

9. Ibid.

10. Ibid., 23.

11. David Hounshell, "The Evolution of Industrial Research in the United States," in *Engines of Innovation: U.S. Industrial Research at the End of an Era*, ed. Richard S. Rosenbloom and William J. Spencer (Boston: Harvard Business School Press, 1996), 50–51.

12. Glen Ross Asner, "The Cold War and American Industrial Research" (PhD diss., Carnegie Mellon University, 2006), 302–24; Ronald Kline, "Construing 'Technology' as 'Applied Science': Public Rhetoric of Scientists and Engineers in the United States, 1880–1945," *Isis* 86, no. 2 (June 1995): 220; Donald E. Stokes, *Pasteur's Quadrant: Basic Science and Technological Innovation* (Washington, DC: Brookings Institution Press, 1997), 55–57.

13. George H. Heilmeier,: An Experiment in Interdisciplinary Research that Worked," *IEEE Transactions on Electron Devices* ED-23, no. 7 (July 1976): 780–85. See also Joshua Shapiro, "George H. Heilmeier," *IEEE Spectrum* 31, no. 6 (June 1994): 56–59; Hedrick Smith, *Rethinking America* (New York: Random House, 1995), 12–18.

14. RCA Laboratories, *Research Report 1968*, 1, David Sarnoff Library Collection, Hagley Museum and Library, Wilmington, DE (DSL).

15. Bilby, *The General*, 280–81. The new wing, renamed the David Sarnoff Library, opened to the public in 1967 and later expanded its holdings to include RCA's technical archive, from which much of the material in this book was drawn. Upon its closure in 2009, these archival collections were transferred to the Hagley Museum and Library in Wilmington, DE.

16. Sobel, *RCA*, 184–86.

17. RCA Laboratories, *Research Report 1968*, 1.

18. Ibid., 1.

19. Quoted in Margaret B. W. Graham, *The Business of Research: RCA and the VideoDisc* (New York: Cambridge University Press, 1986), 86.

20. Ibid., 87–88.

21. Graham, *Business of Research*, 88–89; John van Raalte, interview by author, 3 June 2009, Princeton, NJ.

22. Graham, *Business of Research*, 89. Heilmeier's opposition to IRPCO is based on John van Raalte, interview, 3 June 2009.

23. William Webster, telephone conversation with author, 21 Apr. 2010.

24. RCA Laboratories, *Research Report 1969*, 1, DSL.

25. Heilmeier, "Liquid Crystal Displays," 784.

26. For Zanoni's early LCD work, see Louis Zanoni, Notebook 27707, DSL. For more on the digital voltmeter, see RCA Laboratories, *Progress Report: Liquid Crystals, Oct.–Dec. 1968*, 26; *Progress Report: Liquid Crystals, Jan.–Mar. 1969*, 26, both in DSL. For more on the digital clock, see RCA Laboratories, *Progress Report: Liquid Crystals, Jan.–Mar. 1968*, 20, DSL. For more on the analog clock, see RCA Laboratories, *Progress Report: Liquid Crystals, Apr.–Jun. 1969*, 27, DSL. Lohman's involvement in digital timepiece construction is acknowledged in the Jan.–Mar. 1968 progress report, though his notebooks indicate an interest dating to 1966. See Robert Lohman, Notebook 32487, pp. 9–40, DSL. Hofstein was a member of Lohman's research group whose LCD work will be discussed in more detail later in this chapter.

27. Louis Zanoni, interview by author, 21 Aug. 2009, Ewing, NJ.

28. "Mirror, mirror . . . ," *RCA Radiations* 16, no. 4 (July–Aug. 1969), 2, DSL. See also RCA Laboratories, *Progress Report: Liquid Crystals, Apr.–Jun. 1969*, 27–28; *Progress Report: Liquid Crystals, Jul.–Sep. 1969*, 27, all in DSL.

29. This phrase appeared in the "Objective" section of every LCD progress report between 1967 and 1969.

30. RCA Laboratories, *Progress Report: Liquid Crystals: Oct.–Dec. 1968*, 33, DSL.

31. The contributions of Lechner's group to the Electrofax and Homefax systems are described in the 1968–1969 liquid crystal progress reports as well as J. Tults and D. L. Matthies, "Facsimile Printer Using a Liquid Crystal Array," *RCA Engineer* 17, no. 6 (Apr.–May 1972): 40–43 and Dennis Matthies, interview by author, 19 Aug. 2009, Princeton, NJ.

32. Bernard Lechner, "'Suggested Research Program' on Preliminary Development of a Liquid Crystal Matrix Display," ca. 1969, Bernard Lechner Collection, DSL; "History Crystallized: A First-Person Account of the Development of Matrix-Addressed LCDs for television at RCA in the 1960s," *Information Display* (Jan. 2008): 30.

33. Joseph A. Castellano, *Liquid Gold: The Story of Liquid Crystal Displays and the Creation of an Industry* (Hackensack, NJ: World Scientific, 2005), 67.

34. See, e.g., J. A. Castellano et al., *Field-Effect Liquid Crystal: Final Report* (Jan. 1970), DSL. This report was prepared for the USAF Air Development Center in Rome, NY. An expanded version, dated June 1970 (RADC-TR-70-96), is available via DTIC (Defense Technical Information Center), http://www.dtic.mil/dtic/tr/fulltext/u2/871971.pdf.

35. J. A. Castellano et al., *Liquid Crystal Systems for Electro-Optical Storage Effects*, AFML-TR-73-51, Dec. 1971, http://www.dtic.mil/dtic/tr/fulltext/u2/760173.pdf.

36. J. A. Castellano et al., *Electronically Tuned Optical Filters*, NASA CR-112032, Jan. 1972, DSL; [Project Reports], *Electronically Tuned Optical Filters*, Contract no. NAS-12-638, Apr. 1969, Apr. 1970, DSL. NASA's specific interest in liquid crystals is confirmed in R. H. Hinckley, "Long Range Plan Revision," 20 Aug. 1965,), "Chronological File-August 1965," Box 10, Acc. 255-71A-2309, p. 11, Washington National Records

Center (WNRC), Suitland, MD, and James Elms, "Electronic Research Center Weekly Highlights Report" [Memorandum to NASA Headquarters], 6 June 1968, "Chronological File 1968," Box 3, Acc. 255-71A-2309, pp. 1–3, WNRC.

37. Richard Klein, interview by author, 15 June 2009, Clifton, NJ.

38. *Light Emitting Diode Displays*, Technical Proposal SP-68-47, (Somerville: RCA Electronic Components Division, 21 June 1968), DSL. See also Lawrence A. Murray, Sandor Caplan, and Richard Klein, "Lighting Up in a Group," *Electronics* 41, no. 5 (4. Mar. 1968): 104–10.

39. For more on LED research at RCA during the 1960s and 1970s, see Bob Johnstone, *Brilliant! Shuji Nakamura and the Revolution in Lighting Technology* (Amherst, NY: Prometheus Books, 2007), chap. 2, 4.

40. Klein, interview, 15 June 2009.

41. Ibid.

42. Sandor Caplan, interview by author, 10 June 2009, Princeton, NJ. Caplan dates his involvement with the LCD project to a January 1968 conversation with George Heilmeier at a Society for Information Display meeting in Philadelphia.

43. Caplan, interview, 10 June 2009; Klein, interview, 15 June 2009.

44. Quoted in Hirohisa Kawamoto, "The History of Liquid-Crystal Displays," *Proceedings of the IEEE* 90, no. 4 (Apr. 2002): 482.

45. Klein, interview, 15 June. 2009; Zanoni, interview, 21 Aug. 2009. For DSRC discussions of the benefits of synthetic quartz in LCD construction, see RCA Laboratories, *Progress Report: Liquid Crystals, Oct.–Dec. 1967*, 13; *Progress Report: Liquid Crystals, Apr.–Jun. 1968*, 15, both in DSL.

46. Klein, interview, 15 June 2009.

47. Caplan, interview, 10 June 2009.

48. Klein, interview, 15 June 2009.

49. Klein, interview, 15 June 2009; Castellano, *Liquid Gold*, 64.

50. Klein, interview, 15 June 2009.

51. Caplan, interview, 10 June 2009.

52. Caplan, interview, 10 June 2009. Additional information on the Raritan site comes from the author's interview with semiconductor chemist Norman Goldsmith, 14 Aug. 2009, Princeton, NJ, who worked in both Princeton and Somerville during the course of his career at RCA.

53. Caplan, interview, 10 June 2009.

54. Caplan, interview, 10 June 2009. Heilmeier's contact with Riddel is confirmed in Heilmeier, "Liquid Crystal Displays," and his lab notebooks (e.g., Notebook 39358, p. 34, DSL).

55. George Heilmeier, interview with Margaret Dennis and Carlene Stephens, 4 Dec. 1998, Morristown, NJ.

56. Caplan, interview, 10 June 2009.

57. RCA's R & D expenditures for 1969 are drawn from RCA Corporation, *Annual Report 1970* (New York: RCA, 1971), 26.

58. Caplan, interview, 10 June 2009.

59. Caplan, interview, 10 June 2009; Klein, interview, 15 June 2009. Further information about the day-night mirror can be found in Sandor Caplan, Liquid Crystal Day/Night Mirror, US Patent 3,614,210, filed 6 Nov. 1969 and issued 19 Oct. 1971.

60. Klein, interview, 15 June 2009; Richard I. Klein, Sandor Caplan, and Ralph T. Hansen, Liquid Crystal Display Device, US Patent 3,689,131, filed 29 June 1970 and issued 5 Sept. 1972. For more on DSRC efforts to develop animated LCDs, see Heilmeier, Notebook 39358, p. 34, and Louis A. Zanoni, Color Advertising Display Employing Liquid Crystal, US Patent 3,576,364, filed 20 May 1969 and issued 27 Apr. 1971.

61. Klein, interview, 15 June 2009.

62. Edward J. Homer, *Market Potential of Products Based on Liquid Crystal Dynamic Scattering* (New York: RCA Product and Marketing Planning, 30 Mar. 1970), 111, DSL. According to this report, RCA reached out to Product Planning Associates of Santa Barbara, CA, for assistance because of their "substantial experience in new product development" (3).

63. Homer, *Market Potential*, 111–22.

64. Homer, *Market Potential*, 7–8. The report references meetings with Dr. Glenn Brown (the director of Kent State University's Liquid Crystal Institute) and attendance at trade shows sponsored by IEEE, the Society for Information Display, and the journal *Electro-Optical Systems Design*. It also contained a memo on liquid crystal research at Texas Instruments.

65. Homer, *Market Potential*, 125.

66. The report's timing coincided with Heilmeier and Zanoni's patent for a dynamic scattering display, see George H. Heilmeier and Louis A. Zanoni, Electro-Optical Device, US Patent 3,499,112, filed 31 Mar. 1967 and issued 3 Mar. 1970.

67. Homer, *Market Potential*, 85 (emphasis in original).

68. Ibid., 86.

69. Ibid.

70. Ibid., 85.

71. Gordon E. Moore "Cramming More Components onto Integrated Circuits," *Electronics* 38, no. 8 (19 Apr. 1965): 114–17.

72. James Hillier, "Draft of Remarks by Dr. Hillier At Liquid Crystal Press Conference, 28 May 1968," Bernard Lechner Collection, 6, DSL.

73. Homer, *Market Potential*, 45.

74. Steven Hofstein, Notebook 31938, pp. 1–31, DSL; Solid State Clock, US Patent 3,505,804, filed 23 Apr. 1968 and issued 14 Apr. 1970. This patent describes both clock and watch applications. Although never formally affiliated with the Princeton LCD group, Hofstein discusses his involvement in an oral history conducted 28 Oct. 2011 by

David Laws, Computer History Museum, Mountain View, CA. Transcript available at http://archive.computerhistory.org/resources/access/text/2012/08/102745974-05-01-acc.pdf.

75. Castellano, *Liquid Gold*, 63. For more on RCA's CMOS research, see Ross Knox Bassett, *To the Digital Age: Research Labs, Start-Up Companies, and the Rise of MOS Technology* (Baltimore: Johns Hopkins University Press, 2002), 162–65.

76. Klein, interview, 15 June 2009.

77. Kenyon Kilbon, "Pioneering in Electronics: A Short History of the Origins and Growth of RCA Laboratories, Radio Corporation of America, 1919 to 1964," unpublished typescript, rev. Aug. 1964, 239, http://www.davidsarnoff.org/kil.html.

78. Kilbon, "Pioneering in Electronics," 343. For more on Freedman and his research, see N. S. Freedman, "Superconductive Materials and Magnets: A Review of Progress at RCA," *RCA Engineer* 12, no. 4 (Dec. 1966–Jan. 1967): 7–10.

79. Klein, interview, 15 June 2009.

80. Caplan, interview, 10 June 2009; Klein, interview, 15 June 2009. For more on Lawrence's research, see W. F. Lawrence and A. G. F. Dingwall, "Fossil-Fuel Thermoelectric Generators," *RCA Engineer* 9, no. 3 (Oct.–Nov. 1963): 16–20.

81. RCA Laboratories, *Research Report 1969*, 68.

82. Heilmeier, "Liquid Crystal Displays," 785.

83. Castellano, *Liquid Gold*, 63–64.

84. Homer, *Market Potential*, acknowledgments and p. 7.

85. Graham, *Business of Research*, 71. This characterization aligns with perceived differences between Princeton and Somerville discussed in the author's interviews with Sandor Caplan (10 June 2009), Norman Goldsmith (14 Aug. 2009), Richard Klein (15 June 2009), Alan Sussman (19 Jan. 2011), John van Raalte (10 June 2009), Richard Williams (30 Aug. 2009), and Karl Zaininger (11 Sept. 2009).

86. Castellano, *Liquid Gold*, 57–67; Klein, interview, 15 June. 2009; Zanoni, interview, 21 Aug. 2009.

87. Klein, interview, 15 June 2009.

88. RCA Laboratories, *Progress Report: Liquid Crystals, Apr.–Jun. 1968*, 1; *Progress Report: Liquid Crystals, Jan.–Mar. 1970*,1, both in DSL.

89. Barton became involved with RCA's VideoDisc project beginning in January 1969. (See Lucian Barton, Notebook 39677, pp. 55–56, 66, DSL). He is not listed as a member of the group in the Jan.–Mar. 1969 LCD progress report.

90. Castellano, *Liquid Gold*, 66–67; Zanoni, interview, 21 Aug. 2009; Louis Zanoni and Nunzio (Tony) Luce, interview by Margaret Dennis and Carlene Stephens, 24 Nov. 1998, Trenton, NJ. Optel will be discussed further in chap. 5.

91. John van Raalte, interview, 10 June 2009. Bernard Lechner and two members of his group—Dennis Matthies and Frank Marlowe—expressed similar opinions in their respective interviews with the author on 24 Apr. 2009, 19 Aug. 2009, and 24 Aug. 2009.

92. Homer, *Market Potential*, 89.

93. Graham, *Business of Research*, 40–42.

94. *RCA Liquid Crystals: Domestic Licensing* (New York: [RCA], Sept. 1971), DSL, title page (capitalization in original).

95. Ibid., 16.

96. Ibid., 17 (capitalization in original).

97. Ibid., 2–4.

98. Ibid., 1 (capitalization in original).

99. For more on the development of the Nixie tube, see Larry F. Weber, "Plasma Displays," in *Flat-Panel Displays and CRTs*, ed. Lawrence E. Tannas Jr. (New York: Van Nostrand Reinhold, 1985), 354–55.

100. Sobel, *RCA*, 195–98. See also Alfred D. Chandler Jr., *Inventing the Electronic Century: The Epic Story of the Consumer Electronics and Computer Industries* (New York: Free Press, 2001), 40–44.

101. Johnstone, *Brilliant!*, 92–93.

102. Scott R. Schmedel, "RCA's Plan to Buy Coronet Industries Clears Holders: Minority Objects Loudly," *Wall Street Journal*, 22 Feb. 1971, 6.

103. Ibid.

104. RCA Corporation, *Annual Report 1970* (New York: RCA, 1971), 2. Sarnoff's letter is dated 9 Mar. 1971.

105. Sobel, *RCA*, 175. The manager quoted here is Arthur Beard, who worked as chief engineer for RCA's Computer Systems Division, as confirmed in Franklin M. Fisher, James W. McKie, and Richard B. Mancke, *IBM and the U.S. Data Processing Industry: An Economic History* (New York: Praeger, 1983), 121.

106. Chandler, *Inventing the Electronic Century*, 38–39. For more on RCA and IBM's respective computing programs, see Fisher, McKie, and Mancke, *IBM and the U.S. Data Processing Industry*, 202–13; Kenneth Flamm, *Creating the Computer: Government, Industry, and High Technology* (Washington, DC: Brookings Institution, 1988), 124–25.

107. William D. Smith, "RCA Introduces Four New Computers," *New York Times*, 16 Sept 1970, 69.

108. Sobel, *RCA*, 199–200; Fisher, McKie, and Mancke, *IBM and the U.S. Data Processing Industry*, 215–19.

109. RCA Corporation. "RCA: A Decade of Difference," excerpt from original 16 mm. film, Computer History Museum, http://www.computerhistory.org/revolution/mainframe-computers/7/169/2273.

110. RCA Corporation, *Annual Report 1969* (New York: RCA, 1970), 8–9, 24; , *Annual Report 1970*, 24.

111. Fisher, McKie, and Mancke, *IBM and the U.S. Data Processing Industry*,, 220–24.

112. Sobel, *RCA*, 202–3. The board member quoted here is Martin Seretean, chairman and president of the recently acquired Coronet Industries. Seretean remained one

of Robert Sarnoff's harshest critics before resigning from RCA's board in early 1973. For further information, see George Brown, *And Part of Which I Was: Recollections of a Research Engineer* (Princeton, NJ: Angus Cupar, 1982), 320–23.

113. Quoted in W. David Gardner, "Curtain Act at RCA," *Datamation* (19 Mar. 1972): 41.

114. Gardner's *Datamation* article provides a compelling blow-by-blow reconstruction of RCA's withdrawal from the computer market from the perspectives of company executives and CSD personnel. See also Fisher, McKie, and Mancke, *IBM and the U.S. Data Processing Industry*, 226–27; Sobel, *RCA*, 206–7.

115. Dan Dorfman, "Heard on the Street," *Wall Street Journal*, 5 Oct. 1971, 47.

116. RCA Corporation, *Annual Report 1971* (New York: RCA, 1972), 14.

117. Graham, *Business of Research*, 144–45.

118. Castellano, *Liquid Gold*, 70. The half-dozen figure is based on the roster featured in Joseph Castellano, *Liquid Crystal Research Program: 1972 Annual Report*, Jan. 1973, 1, Joseph Castellano Collection, DSL. It does not include consultants at the DSRC or Somerville.

119. Klein, interview, 15 June 2009.

120. Klein, interview, 15 June 2009. For confirmation of the LCD operation's return to Somerville, see Lawrence Goodman, interview by author, 21 Aug. 2009, Princeton, NJ; H. C. Schindler, "Liquid-Crystal Dynamic Scattering for Display Devices," *RCA Engineer* 17, no. 6 (Apr.–May 1972): 35–39. For more on ECD's name change, see "New Solid-State Division Formed," *RCA Engineer* 15, no. 6 (Apr.–May 1970): 91.

121. Caplan, interview, 10 June 2009; Klein, interview , 15 June 2009.

122. RCA Corporation, *Annual Report 1971*, 15. This report notes that RCA employed 131,000 people at the end of 1970 and only 118,000 people by the end of 1971.

123. Cade Metz, "Liquid Crystals-Display Genius No Match for Petty Politics," *Register*, 27 Apr. 2009, http://www.theregister.co.uk/2009/04/27/george_heilmeier/.

124. Kyoto Prize, "2005: Advanced Technology: Electronics: George H. Heilmeier," KyotoPrize.org, www.kyotoprize.org/en/laureates/george_h_heilmeier/; "Engineering News and Highlights," *RCA Engineer* 15, no. 1 (June–July 1969): 95.

125. Heilmeier, quoted in Metz, "Liquid Crystals-Display Genius." Heilmeier also describes RCA's opposition to forming a spin-off company in his interview with Maggie Dennis and Carlene Stephens, 4 Dec. 1998.

126. George Heilmeier, interview by Stewart Finley, *Men and Molecules*, episode 515 ("Liquid Crystals: A Bright Promise"), ca. 1970.

127. "Engineering News and Highlights," *RCA Engineer* 20, no. 2 (Aug.–Sept. 1974): 95.

128. For more on the variable relationship between RCA's research and semiconductor divisions, see Hyungsub Choi, "The Boundaries of Industrial Research: Making Transistors at RCA, 1948–1960," *Technology and Culture* 48, no. 4 (Oct. 2007): 758–82.

129. Heilmeier, "Liquid Crystal Displays," 785.

130. Ibid.

131. Ibid.

132. Clayton M. Christensen, *The Innovator's Dilemma* (New York: Harper Business, 2011), xviii.

133. Christensen, *Innovator's Dilemma*, xxiv–xxv, chap. 6.

134. See, e.g., Richard Klein, interview, 15 June 2009; Alan Sussman, interview, 19 Jan. 2011; Zanoni, interview, 21 Aug. 2009.

Chapter Five

1. Kenneth Bilby, *The General: David Sarnoff and the Rise of the Communications Industry* (New York: Harper and Row, 1986), 279–85.

2. Ibid., 283.

3. "David Sarnoff of RCA is Dead: Visionary Broadcast Pioneer," *New York Times*, 13 Dec. 1971, 1; "Television This Week," *New York Times*, 12 Dec. 1971, D25.

4. Rockefeller's eulogy contrasts sharply with more recent popular biographies of Sarnoff, which emphasize his battles with television pioneer Philo Farnsworth and FM inventor Edwin Armstrong. See, e.g., Thomas S. W. Lewis, *Empire of the Air: The Men Who Made Radio* (New York: HarperCollins, 1991); Daniel Stashower, *The Boy Genius and the Mogul: The Untold Story of Television* (New York: Broadway Books, 2002); Scott Woolley, *The Network: The Battle for the Airwaves and the Birth of the Communications Age* (New York: HarperCollins, 2016).

5. *Eulogies, David Sarnoff 1891–1971*, David E. Lilienthal Papers, Box 498: 1971 Correspondence: Name Files (S-Z), RE: Sarnoff, David 1971, Public Policy Papers, p. 2, Department of Rare Books and Special Collections, Princeton University Library, Princeton, NJ. See also Laurie Johnston, "700 at Sarnoff Service in Emanu-El," *New York Times*, 16 Dec. 1971, 38.

6. "General David Sarnoff, 1891–1971," *RCA Radiations* 18 (Winter 1971–1972), 3, David Sarnoff Library Collection, Hagley Museum and Library, Wilmington, DE (DSL).

7. RCA Laboratories, *Research Report 1971*, 1, DSL.

8. Ibid., 1.

9. RCA Laboratories, *Research Report 1971*, 1, 77–84. The definitive discussion of RCA's home video player projects is Margaret B. W. Graham, *The Business of Research: RCA and the VideoDisc* (New York: Cambridge University Press, 1986).

10. Graham, *Business of Research*, chaps. 5–7.

11. Clayton M. Christensen, *The Innovator's Dilemma* (New York: Harper Business, 2011), 139–40, 165–66.

12. Ibid., 147–49, 176–78, 191–93.

13. Joseph Castellano discusses three of these LCD spin-offs—Optel, Princeton

Materials Science, and Ashley-Butler—in *Liquid Gold: The Story of Liquid Crystal Displays and the Creation of an Industry* (Hackensack, NJ: World Scientific, 2005), 91–100. For discussions of spin-offs in the semiconductor industry, see Christophe Lécuyer, *Making Silicon Valley: Innovation and the Growth of High Tech, 1930–1970* (Cambridge, MA: MIT Press, 2006), chaps. 6–7.

14. RCA Laboratories, *Research Report 1960*, 66, DSL. See also Bruce Shore, "Lasers: New Power from Light," *Electronic Age* 21, no. 3 (Summer 1962): 8–11.

15. RCA Laboratories, *Research Report 1960*, 66–67.

16. Zoltan Kiss, interview by author, 5 Sept. 2013, Belle Mead, NJ.

17. Ibid.

18. Kiss, interview, 5 Sept. 2013; Shore, *Lasers*, 10–11.

19. Kiss, interview, 5 Sept. 2013; Shore, *Lasers*, 10–11. See also Donald S. McClure and Zoltan Kiss, "Survey of the Spectra of the Divalent Rare-Earth Ions in Cubic Crystals," *Journal of Chemical Physics* 39, no. 12 (15 Dec. 1963): 3251–57.

20. H. R. Lewis, "The Laser: An Introduction," *RCA Engineer* 8, no. 5 (Feb.–Mar. 1963): 6–7.

21. RCA Laboratories, *Research Report 1962*, 66, DSL. See also R. C. Duncan Jr., Z. J. Kiss, and H. R. Lewis, "Sun-Pumped Continuous Laser," *RCA Engineer* 8, no. 5 (Feb.–Mar. 1963): 62.

22. Kiss, interview, 5 Sept. 2013.

23. "Science Notes: New Laser," *New York Times*, 21 Oct. 1962, E7.

24. Bruce Shore, "'Op' Electronics," *Electronic Age* 27, no. 4 (Autumn 1968): 13–14.

25. Kiss, interview, 5 Sept. 2013. See also Shore, "'Op' Electronics," 14.

26. Zoltan J. Kiss, "Photochromics," *Physics Today* 23, no. 1 (Jan. 1970): 42–49; RCA Laboratories, *Research Report 1965*, 171, DSL.

27. RCA Laboratories, *Research Report 1967*, 109–10, DSL. See also Kiss, "Photochromics," 49; William Phillips and Zoltan J. Kiss, "Photo Erasable Dark Trace Cathode-Ray Storage Tube," *Proceedings of the IEEE* 56, no. 11 (Nov. 1968): 2072–73; Shore, "'Op' Electronics," 14.

28. RCA Laboratories, *Research Report 1967*, 101–2. See also H. R. Lewis, "Quantum Electronics Research at RCA Laboratories: An Introduction," *RCA Engineer* 12, no. 3 (Oct.–Nov. 1966): 24.

29. Webster and Lewis' respective promotions are discussed in: "Engineering News and Highlights," *RCA Engineer* 14, no. 3 (Oct.–Nov. 1968): 94; "Engineering News and Highlights," *RCA Engineer* 14, no. 4 (Dec. 1968–Jan. 1969): 95.

30. Kiss, interview, 5 Sept. 2013.

31. Ibid.

32. Charles G. Burck, "Optel's (Mis)adventures in Liquid Crystals," *Fortune*, Oct. 1973, 194.

33. Although people occasionally referred to *Silicon Valley* in the late 1960s, the first

use of the term in print was Donald Hoefler's three-part article "Silicon Valley, USA," published by *Electronic News* in January 1971. See Arnold Thackray, David C. Brock, and Rachel Jones, *Moore's Law: The Life of Gordon Moore, Silicon Valley's Quiet Revolutionary* (New York: Basic Books, 2015), 319.

34. Kiss, interview, 5 Sept. 2013.

35. Castellano, *Liquid Gold*, 92; Edward Kornstein, telephone conversation with author, 17 Sept. 2014; Zoltan Kiss, interview by author, 20 Sept. 2013, Belle Mead, NJ.

36. *Zoltan J. Kiss, RCA Laboratories, 1960–1969* [7 Nov. 1969], Zoltan Kiss Personal Archive, Belle Mead, NJ (KPA).

37. Robert L. Quinn, "Zoltan's Last Supper," cartoon dated 7 Nov. 1969, KPA.

38. P. E. Seeley, "Electro-Optical Engineering at Burlington," *RCA Engineer* 9, no. 6 (Apr.–May 1964): 63–67.

39. Kornstein, telephone conversation, 17 Sept. 2014; Louis Zanoni and Nunzio (Tony) Luce, interview by Margaret Dennis and Carlene Stephens, 24 Nov. 1998, Trenton, NJ.

40. Kiss, interview, 20 Sept. 2013.

41. Ibid.

42. Burck, "Optel's (Mis)adventures," 196.

43. Optel Corporation, *Optel: Pioneers in Electronic Digital Watch Technology* [promotional brochure] (Princeton: Optel, ca. 1975), 1, KPA; Zanoni and Luce, interview, 24 Nov. 1998; Kornstein, telephone conversation, 17 Sept 2014. See also Castellano, *Liquid Gold*, 92.

44. Quantel Corporation, "Technical Proposal No. 700601-1: Development Program of a Liquid Crystal Watch Display for Bulova Watch Company," 1 June 1970. Drawn from files on Refac Intl Ltd v. Hitachi Ltd et al., 2:87-cv-06191-TJH, Accession 021-96-0107 Location 20015511, Box 194-202, 8, National Archives at Riverside, Perris, CA. (Thanks to George Curley of the University of California, Riverside for assistance retrieving this document.)

45. Scott G. Knowles and Stuart W. Leslie, "'Industrial Versailles': Eero Saarinen's Corporate Campuses for GM, IBM and AT&T," *Isis* 92, no. 1 (Mar. 2001): 1–33. For more on US government policies encouraging the construction of separate research laboratories, see Glen Ross Asner, "The Cold War and American Industrial Research" (PhD diss., Carnegie Mellon University, 2006), chap. 4.

46. Ross Knox Bassett, *To the Digital Age: Research Labs, Start-Up Companies, and the Rise of MOS Technology* (Baltimore: Johns Hopkins Univ. Press, 2002), 182, 284–87.

47. Kiss, interview, 20 Sept. 2013; Burck, "Optel's (Mis)adventures," 196.

48. Louis Zanoni, interview by author, 21 Aug. 2009, Ewing, NJ.

49. Zanoni and Luce, interview, 24 Nov. 1998.

50. Optel Corporation, "Optel Reflicon Display Model D-10," 15 Mar. 1971, KPA; "Optel Terminal Model T710," ca. 1971, KPA.

51. Zanoni, interview, 21 Aug. 2009. See also Amilcar Guimaraes, "The Beginning of the Liquid Crystal Watch," *National Association of Watch and Clock Collectors Bulletin* 321 (Aug. 1999): 479–83.

52. Zanoni and Luce, interview, 24 Nov. 1998. Staff numbers derived from Quantel Corporation, "Technical Proposal No. 700601-1," 8.

53. Zanoni and Luce, interview 24 Nov. 1998.

54. Ibid.

55. "Making Displays Easier to Read," *Business Week*, 21 Nov. 1970, 49; E. C. Terry, "Optel Challenging CRTs with Liquid Crystal Entry," *Electronic News*, 16 Nov. 1970, 62.

56. Kornstein, telephone conversation, 17 Sept. 2014.

57. Zanoni and Luce, interview, 24 Nov. 1998.

58. Burck, "Optel's (Mis)adventures," 196, 201. The other directors were Lawrence Goldmuntz and Edwin Robbins. For a Swiss perspective on this story, see Claude-Pierre Chambet, "La petite histoire des cristaux liquides," *FAN-L'Express*, 2–3 Feb. 1974, 7.

59. Zanoni and Luce, interview, 24 Nov. 1998.

60. Zanoni, interview, 21 Aug. 2009. For an account of this management struggle from the perspective of Kiss's opposition, see Douglas Bosomworth, Lawrence A. Goldmuntz, Edward Kornstein, and Edwin Robbins, "Letter to the Stockholders of Quantel Corporation," 1 Dec. 1970, Louis Zanoni Personal Archive, Ewing, NJ (ZPA).

61. Quoted in Burck, "Optel's (Mis)adventures," 196.

62. Kiss, interview, 20 Sept. 2013.

63. Zanoni and Luce, interview, 24 Nov. 1998.

64. Ibid.

65. Quantel Corporation, "Technical Proposal No. 700601-1," summary.

66. Carlene Stephens and Maggie Dennis, "Engineering Time: Inventing the Electronic Wristwatch," *British Journal for the History of Science* 33, no. 4 (Dec. 2000): 480–82.

67. Stephens and Dennis, "Engineering Time," 492–94; Castellano, *Liquid Gold*, 108.

68. Quantel Corporation, "Technical Proposal No. 700601-1," 1.

69. Zanoni and Luce, interview, 24 Nov. 1998.

70. Ibid.

71. Zanoni, interview, 21 Aug. 2009.

72. Zanoni and Luce, interview, 24 Nov. 1998. For further details on the watch's construction, see Nunzio A. Luce, "C/MOS Digital Watch Features Liquid Crystal Display," *Electronics*, 10 Apr. 1972, 93–97; Guimaraes, "Beginning of the Liquid Crystal Watch."

73. Zanoni and Luce, interview, 24 Nov. 1998.

74. Kiss, interview, 20 Sept. 2013. See also Stephens and Dennis, "Engineering Time," 485–94; "Swiss Watchmakers Buy American," *Business Week*, 18 Mar. 1972, 21.

75. Luce, "C/MOS Digital Watch," 93.

76. Bassett, *To the Digital Age*, 161–65.

77. Stephens and Dennis, "Engineering Time," 493.

78. Zanoni and Luce, interview, 24 Nov. 1998.

79. Zanoni and Luce, interview, 24 Nov. 1998; Kornstein, telephone conversation, 17 Sept. 2014.

80. Kiss, interview, 20 Sept. 2013; Kornstein, telephone conversation, 17 Sept. 2014.

81. Zanoni and Luce, interview, 24 Nov. 1998. "Bulova to Introduce Quartz Crystal Watch That Retails for $395," *Wall Street Journal*, 6 Dec. 1971, 17.

82. "L''octoscope' ouvre la voie à la montre 100% électronique," *FAN-L'Express*, 15 Apr. 1971, 3. For more on Optel's ties with Omega, including a proposal to design liquid crystal watch displays, see Zoltan J. Kiss, Zoltan J. Kiss to Dr. Hans Widmer [Omega], Letter dated 3 Mar. 1971, ZPA.

83. "Swiss watchmakers buy American," 21; Claude-Pierre Chambet, "Coup de maître! La Société des Garde-Temps va commercialiser, en première mondiale, une montre à quartz à affichage digital par cristaux liquides," *FAN-L'Express*, 7 Mar. 1972, 3. For more on SGT, see Lucien F. Trueb, Günther Ramm, and Peter Wenzig, *Electrifying the Wristwatch* (Atglen, PA: Schiffer, 2013), 186–87.

84. Kiss, interview, 20 Sept. 2013.

85. Ibid.

86. Chambet, "Coup de maître!," 3; Gil Baillod, "SGT prêt à commercialiser une montre sans rouages," *L'Impartial*, 7 Mar. 1972, 1.

87. Baillod, "SGT prêt à commercialiser," 3.

88. Claude-Pierre Chambet, "Un nom: Ditronic mais ni roue ni aiguille . . . ," *FAN-L'Express*, 16 Mar. 1972, 3. See also "Encore une 'Digitale' d'Optel . . . ," *FAN-L'Express*, 23 Apr. 1972, 7; "Electronic Watches for the Mass Market," *Business Week*, 22 Apr. 1972, 62, 66; "Swiss Watchmakers Buy American," 21.

89. "Swiss watchmakers buy American," 21.

90. "Optel Corp. Offering Sells Out," *Wall Street Journal*, 22 June 1972, 29.

91. Joseph A. Castellano, "Liquid Crystals for Electro-Optical Application," *RCA Review* 33, no. 1 (Mar. 1972): 296–310. *RCA Review* was a quarterly technical journal published by RCA Laboratories.

92. Ibid., 299–308. This article also contained a discussion of "field-induced phase changes," which is described later in this section.

93. Ibid., 309–10.

94. M. Schadt and W. Helfrich, "Voltage-Dependent Optical Activity of a Twisted Nematic Liquid Crystal," *Applied Physics Letters* 18, no. 4 (15 Feb. 1971): 127–28.

95. Castellano, *Liquid Gold*, 72.

96. Wolfgang Helfrich, interview by author, 28 Sept. 2009, Berlin, Germany. For examples of Helfrich's research during this period, see W. Helfrich, *A Molecular Theory of Flow Alignment of Nematic Liquid Crystals*, PTR-2473, 24 May 1968; *Alignment Inversion Walls in Nematic Liquid Crystals in the Presence of a Magnetic Field*, PTR-2567, 12 Sept.

1968; *Conduction-Induced Alignment in Nematic Liquid Crystals: Basic Model and Stability Considerations*, PTR-2686, 26 Feb. 1969, all in DSL.

97. Helfrich, interview, 28 Sept. 2009.

98. Charles Mauguin, "Sur les cristaux liquides des Lehmann," *Bulletin de la Société française de Minéralogie* 34 (1911): 71–117, translated as "On the Liquid Crystals of Lehmann" in *Crystals That Flow: Classic Papers from the History of Liquid Crystals*, ed. Timothy J. Sluckin, David A. Dunmur, and Horst Stegemeyer (New York: Taylor and Francis, 2004), 100–121.

99. In technical terms, Helfrich's concept relied on the use of nematic materials such as PEBAB (p-ethoxybenzylidene-p-aminobenzonitrile), which exhibited positive dielectric anisotropy. Castellano, *Liquid Gold*, 72, confirms that Helfrich was aware of such materials, which had been the subject of experiments at the DSRC.

100. Helfrich, interview, 28 Sept. 2009.

101. Hirohisa Kawamoto, "The History of Liquid-Crystal Displays," *Proceedings of the IEEE* 90, no. 4 (Apr. 2002): 473.

102. See, e.g., W. Helfrich, *Capillary Viscometry of Cholesteric Liquid Crystals*, PEM-3187, 20 Oct. 1969; *Capillary Flow of Cholesteric and Smectic Liquid Crystals*, PTR-2736, 15 May 1969; *Deformation of Cholesteric Liquid Crystals with Low Threshold Voltage*, PRRL-70-TR-164, 12 Aug. 1970; *Electrohydrodynamic and Dielectric Instabilities of Cholesteric Liquid Crystals*, PRRL-70-TR-236, 4 Nov. 1970, all in DSL.

103. RCA Laboratories, *Progress Report: Liquid Crystals, Apr.–Jun. 1969*, 14–17, DSL; Castellano, "Liquid Crystals in Electro-Optical Application," 307–8. Heilmeier first described the possibility of a display using field-induced phase change in cholesteric crystals in September 1965 (Heilmeier, Notebook 28088, pp. 56–62). By 1969, he returned to the idea, using new cholesteric-nematic mixtures. For further information, see George H. Heilmeier and Joel E. Goldmacher, "Electric-Field-Induced Cholesteric-Nematic Phase Change in Liquid Crystals," *Journal of Chemical Physics* 51, no. 3 (Aug. 1969): 1258–60.

104. Castellano, *Liquid Gold*, 72. In oral history interviews with the author, Lawrence Goodman (21 Aug. 2009), Alan Sussman (19 Jan. 2011), and Louis Zanoni (21 Aug. 2009) also suggested that Helfrich may have been working on twisted nematic liquid crystals while at the DSRC.

105. J. A. Castellano et al., *Electronically Tuned Optical Filters*, NASA CR-112032, Jan. 1972; *Electronically Tuned Optical Filters*, Contract No. NAS-12-638, Apr. 1969, Apr. 1970, all in DSL.

106. The former explanation for Heilmeier's actions is presented in David Dunmur and Tim Sluckin, *Soap, Science and Flat Screen TVs: A History of Liquid Crystals* (New York: Oxford University Press, 2011 [2014]), 204–5.

107. Helfrich, interview, 28 Sept. 2009.

108. Castellano, *Liquid Gold*, 73–74. For a detailed firsthand account of Helfrich and Schadt's LCD experiments, see Martin Schadt, "Milestone in the History of Field-Effect Liquid Crystal Displays and Materials," *Japanese Journal of Applied Physics* 48 (2009), doi:org/10.1143/JJAP.48.03B001.

109. Dunmur and Sluckin, *Soap, Science and Flat Screen TVs*, 217–19; Schadt and Helfrich, "Voltage-Dependent Optical Activity."

110. Kawamoto, "History of Liquid-Crystal Displays," 471–73.

111. Schadt, "Milestone," 4; Castellano, *Liquid Gold*, 75–76.

112. James Fergason, Display Devices Utilizing Liquid Crystal Light Modulation, US Patent 3,731,986, filed 22 Apr. 1971 and issued 8 May 1973; W. Helfrich and M. Schadt, Optical Device, Swiss Patent 532,261, filed 4 Dec. 1970 and issued 15 Feb. 1972. The two-month figure is based on the fact that Fergason's April 1971 patent filing was the continuation of a copending application (Ser. No. 113,948) filed 9 Feb. 1971.

113. Dunmur and Sluckin, *Soap, Science and Flat Screen TVs*, 217–19. For a discussion of this legal dispute written from Fergason's perspective, see Terri Fergason Neal and Marian Pierce, *Mr. Liquid Crystal: The Untold Story of How James L. Fergason Invented the Liquid Crystal Display and Helped Create the Digital World* (Los Angeles: New Insights Press, 2016), chap. 16.

114. Castellano, *Liquid Gold*, 67–70

115. H. C. Schindler, "Liquid-Crystal Dynamic Scattering for Display Devices," *RCA Engineer* 17, no. 6 (Apr.–May 1972): 35–39.

116. Barry Kramer, "Chemical Chameleons: Liquid Crystals, Once Merely Curiosities, May Become Boon to Industry, Consumers," *Wall Street Journal*, 29 Feb. 1972, 36.

117. Joseph A. Castellano, Lawrence A. Goodman, and Ronald N. Friel, *Human Factor Analysis of Reflective Liquid Crystal Displays*, PRRL-72-TR-137, 30 Aug. 1972, 14, DSL. The study's June 1972 start date is confirmed in Joseph Castellano, Notebook 45534, DSL, p. 37.

118. Castellano et al., *Human Factor Analysis*, 13–14.

119. Joseph A. Castellano, *Liquid Crystal Research Program: 1972 Annual Report*, Jan. 1973, p. 1, Joseph Castellano Collection, DSL.

120. RCA Laboratories, *Research Report 1972*, 118–19, DSL. Additional information on RCA's liquid crystal research during this period can be found in Joseph Castellano, *Liquid Crystal Research Program: Progress Report, Jan.–Feb. 1972*; E. F. Hockings, *Glass and Organic Materials Research: Quarterly Progress Report, Dec. 1971–Feb. 1972, Mar.–May 1972, Sep.–Nov. 1972*, all in Joseph Castellano Collection, DSL.

121. Burck, "Optel's (Mis)adventures," 202; Castellano, *Liquid Gold*, 84–85. Seiko's LCD wristwatch is discussed further in this book's conclusion.

122. RCA Laboratories, *Research Report 1973*, 130, DSL.

123. "RCA Building New Jersey Plant," *Wall Street Journal*, 21 Feb. 1974, 8.

124. Stacy V. Jones, "New Liquid Crystal Is Devised," *New York Times*, 23 Feb. 1974,

39; Bill D. Ross, "A Bank Loan for 2d-Graders," *New York Times*, 12 May 1974, 79. The October 1974 opening date is listed in RCA Laboratories, *Research Report 1974*, 180, DSL.

125. Sussman, interview by author, 19 Jan. 2011, Princeton, NJ. The timing of Sussman's departure from Princeton is based on the final entry in his laboratory notebook: Alan Sussman, Notebook 46488, DSL, p. 24.

126. Sussman, interview, 19 Jan. 2011. Because the glass had to be held at specific angles as the films were deposited in a gaseous state, this process was known as "sloped evaporation." It was not an RCA invention—see Castellano, *Liquid Gold*, 115—but Sussman appears to have overseen its implementation within the company's LCD manufacturing operation.

127. Kawamoto, "History of Liquid-Crystal Displays," 479; Sussman, interview, 19 Jan. 2011. This patchy appearance resulted from the equal likelihood that the molecules in a twisted nematic LCD would retwist in a clockwise or counterclockwise direction following the removal of an electric field. Sussman's solution to this "reverse twist" problem essentially made it more likely that the molecules would orient themselves the same way. He first published the idea in 1972 during a brief stint at Optel, but Kawamoto suggests the idea predated his firing from RCA. See Alan Sussman, "Electrooptic Liquid Crystal Devices: Principles and Applications," *IEEE Transactions on Parts, Hybrids, and Packaging* PHP-8, no. 4 (Dec. 1972): 24–37.

128. Kawamoto, "History of Liquid-Crystal Displays," 497.

129. Farina is something of an enigma. He is implied to be in charge of SSD's Liquid Crystal Engineering program in the Apr.–May 1973 issue of *RCA Engineer* (p. 84), but his formal appointment as Director of Liquid Crystal Operations is not mentioned until two years later (*RCA Engineer*, Apr.–May 1975, 92). Despite limited ties with the Princeton LCD group, his long-standing interest in electronic displays is confirmed in: P. L. Farina, "Digital Readouts," *RCA Engineer* 17, no. 6 (Apr.–May 1972): 12–16.

130. Sussman, interview, 19 Jan. 2011.

131. Lawrence Goodman, interview by author, 21 Aug. 2009, Princeton, NJ.

132. Castellano, *Liquid Gold*, 132–34; Kathleen McDermott, *Timex: A Company and Its Community, 1854–1998* (New York: Timex Corporation, 1998), 182–83, 192–93; Martin Gold, "LCDs, Semicons Involved in Timex Deal with RCA," *Electronic News*, 16 Feb. 1976, 63.

133. Statistics derived from *Moody's OTC Industrial Manual* (New York: Moody's Investors Service, 1972–1979).

134. Burck, "Optel's (Mis)adventures," 201; "Swiss Watchmakers Buy American," 21; "Electronic Watches for the Mass Market," 62.

135. "Swiss Watchmakers Buy American," 21.

136. Burck, "Optel's (Mis)adventures," 201; Kornstein, telephone conversation, 17 Sept. 2014.

137. Optel Corporation, *Interim Report for the Three Months Ended March 31, 1973*, p. 1, ZPA.

138. Burck, "Optel's (Mis)adventures," 201–2 Kornstein, telephone conversation, 17 Sept. 2014; "Watch Industry Shifting to Electronic Movements," *Christian Science Monitor*, 21 Aug. 1973, 6.

139. Nat Snyderman, "Need Improvements: Optel," *Electronic News*, 19 Nov. 1973, 49.

140. Snyderman, "Need Improvements," 49. Barnett's background is discussed in "Optel to Develop Auto Displays," *Electronic News*, 15 Oct. 1973, 52.

141. Kiss, interview by author, 20 Sept. 2013, Belle Mead, NJ; Burck, "Optel's (Mis) adventures," 202.

142. Optel Corporation, *Annual Report 1973* (Princeton: Optel, 1974), 1, ZPA; Snyderman, "Need Improvements," 49.

143. "Optel Corp. to Report Loss for Quarter, Year," *Wall Street Journal*, 11 Jan. 1974, 4.

144. Zanoni and Luce, interview, 24 Nov. 1998.

145. Kornstein, telephone conversation, 17 Sept. 2014.

146. For the terms of this arrangement between Optel and American Express, see Ramiro Medina to David J. Barnette [*sic*], letter dated 6 Nov. 1974, ZPA. Medina was the assistant to American Express's vice president for merchandise sales.

147. American Express Company, "Introducing the Quartz Segtronic," Mail-order brochure, ca. 1975, 1, ZPA.

148. Zanoni and Luce, interview, 24 Nov. 1998.

149. Kornstein, telephone conversation, 17 Sept. 2014.

150. "Optel Corp. Said It Obtained Bank Credit of $1 Million," *Wall Street Journal*, 28 Mar. 1974, 23.

151. Martin Gold, "Lewis Appointed Optel President," *Electronic News*, 8 Apr. 1974, 1.

152. Martin Gold, "ILC's Heller Heads Optel: Kiss Stays as Chairman," *Electronic News*, 9 Feb. 1976, 46; Zanoni and Luce, interview, 24 Nov. 1998.

153. "Optel Gets $1.5 Million Loan," *Wall Street Journal*, 26 July 1974, 17; "Optel Gets Mitsubishi Loan," *Electronic News*, 29 July 1974, 17.

154. *Optel: Pioneers in Electronic Digital Watch Technology*, 12.

155. Zanoni and Luce, interview, 24 Nov. 1998. For broader industry perspectives on LED vs. LCD readouts, see "Optoelectronics," *Electronic News*, 9 Feb. 1976, 1, 50–51, 54.

156. Zanoni and Luce, interview, 24 Nov. 1998.

157. American Express Company, "Introducing the Ultimatic for men. And the Ultimatic for women," Mail-order brochure, ca. 1975, ZPA.

158. Zanoni and Luce, interview, 24 Nov. 1998.

159. "Calculator/Watch Fits on the Wrist," *Electronics*, 1 May 1975, 40–42; Nick Poulos, "New Watches Will Tell More Than Time," *Chicago Tribune*, 20 Aug. 1975, C7.

160. "Optel Corp. Elects Six to Board, Completes Stock-Debt Exchange," *Wall Street*

Journal, 24 Dec. 1975, 10; "A Battle against Time Is the Test at Optel," *Business Week*, 19 July 1976, 22–23.

161. Gold, "ILC's Heller Heads Optel," 46.

162. Ibid.

163. Gold, "LCDs, Semicons Involved in Timex Deal with RCA," 63.

164. Castellano, *Liquid Gold*,70, 96–97.

165. Graham, Business of Research, 145.

166. RCA Laboratories, *Research Report* 1972, 1.

167. Ibid.

168. "RCA Says Earnings Rising in 2nd Period: Full-Year High Possible," *Wall Street Journal*, 2 May 1973, 23. See also "RCA Says New Product Development Might Make '73 a Record Profit Year," *Los Angeles Times*, 2 May 1973, F9.

169. "Where the Company Goes from Here," *Communicate, The Magazine of RCA*, July–Aug. 1973, 14, DSL.

170. "Ibid.

171. John van Raalte, *RCA Laboratories' Research Program on Large-Area Flat TV Displays*, Nov. 1973, 1, DSL. The timing of the Task Force's formation is found in RCA Laboratories, *Research Report* 1972, 101.

172. Van Raalte, *RCA Laboratories' Research Program*, 1–2.

173. Bob Johnstone, *Brilliant! Shuji Nakamura and the Revolution in Lighting Technology* (Amherst, NY: Prometheus Books, 2007), 85–101; Christophe Lécuyer and Takahiro Ueyama, "The Logics of Materials Innovation: The Case of Gallium Nitride and Blue Light Emitting Diodes," *Historical Studies in the Natural Sciences* 43, no. 3 (June 2013): 243–80.

174. Van Raalte, *RCA Laboratories' Research Program*, 10–13. For more on the status of electroluminescent display research in the 1970s, see B. Kazan, "Electroluminescent Displays," *Proceedings of the SID* 17, no. 1 (1976): 23–29.

175. Van Raalte, *RCA Laboratories' Research Program*, 13–15. In addition to LCDs, the report also touched on electrophoretic and electrochromic displays. All three are discussed in L. A. Goodman, "Passive Liquid Displays: Liquid Crystals, Electrophoretics, and Electrochromics," *IEEE Transactions on Consumer Electronics* CE-21, no. 3 (Aug. 1975): 247–59.

176. Goodman, interview, 21 Aug. 2009.

177. Van Raalte, *RCA Laboratories' Research Program*, 2, 13. There is some ambiguity in this report concerning the number of picture elements in the final display. The introduction specifies a resolution of 250,000 (= 500 × 500) elements, but the section on liquid crystals argues that "a diode or transistor must be incorporated in the panel for each picture element ($\geq 5 \times 10^5$)."

178. Van Raalte, *RCA Laboratories' Research Program*, 13; Goodman, interview, 21 Aug. 2009.

179. Van Raalte, *RCA Laboratories' Research Program*, 16.

180. Ibid.

181. Ibid., 19.

182. John van Raalte, interview by author, 10 June 2009, Princeton, NJ; T. L. Credelle, "Large-Screen Flat-Panel Television: A New Approach," *RCA Engineer* 26, no. 7 (July–Aug. 1981): 75–81.

183. RCA Laboratories, *Research Report 1973*, 104, DSL. Joseph Castellano confirmed Ross's leadership role in an e-mail message to the author, 31 Jan. 2011. Goodman, interview, 21 Aug. 2009.

184. RCA Laboratories, *Research Report 1973*, 130; *Research Report 1974* 135–36; *Research Report 1975*, 44–45, DSL.

185. RCA Corporation, *Annual Report 1974* (New York: RCA, 1975), 19; Stacy V. Jones, "Testing System Is Found for Integrated Circuits," *New York Times*, 24 Jan. 1976, 37. Channin discussed this work in his interview with the author, 17 Aug. 2009, Princeton, NJ.

186. Gold, "LCDs, Semicons in Timex Deal with RCA," 1.

187. Gene Smith, "Sarnoff's Departure Is Called Result of RCA 'Palace Revolt,'" *New York Times*, 7 Nov. 1975, 49.

188. RCA Corporation, *Annual Report 1975* (New York: RCA, 1976), 6.

189. "Agreement between Timex Corporation and RCA Corporation for Purchase and Sale of RCA Corporation's Liquid Crystal Display Operations," 31 Mar. 1976. The closing date (12 Apr. 1976) is confirmed in section 3.01 (p. 5), the purchase price can be found in section 2.03 (p. 3–4), and the details concerning RCA's intellectual property portfolio are discussed in section 4.14 (p. 13–15) and Exhibit K (p. K1–K11). The author is indebted to Lou Galie, Timex Group's senior vice president in charge of technology, for supplying a copy of this document.

190. "Kiss Resigns as Optel Chairman," *Electronic News*, 26 Apr. 1976, 26; Robert D. Hershey Jr., "Solar Power Race at Jersey Concern," *New York Times*, 2 July 1982, D1.

191. "3 Leave Optel; Kiss May Follow," *Electronic News*, 22 Mar. 1976, 34.

192. "Optel Says It's Seeking an Infusion of Capital to Ensure Its Survival," *Wall Street Journal*, 16 Apr. 1976, 21.

193. Gil Baillod, "Horlogerie: Un Américain (Optel) à bout de souffle." *L'Impartial*, 27 Apr. 1976, 11; Claude-Pierre Chambet, "Si Optel ne paie pas ce qu'il doit à un industriel jurassien son stand risque d'être fermé ce matin à la Foire de Bâle!" *FAN-L'Express*, 28 Apr. 1976, 11.

194. Gil Baillod, "Optel demande un moratoire," *L'Impartial*, 25 June 1976, 11; "Venez à New York le 30 juin! OPTEL dépasse les borne à moins que . . . ," *FAN-L'Express*, 22 June 1976, 17.

195. "Optel Says Demands from Creditors Force It to File for Chapter 11," *Wall Street Journal*, 17 June 1976, 10; "En attendant qu'Optel trouve de l'argent, ses créanciers

suisses lui accordent un sursis jusqu'au 6 juin . . . ," *FAN-L'Express*, 8–9 May 1976, 9; "A Battle against Time Is the Test at Optel," 23.

196. RCA Corporation, *Annual Report 1976* (New York: RCA, 1977), 1–2.

197. Channin, interview, 17 Aug. 2009; Goodman, interview, 21 Aug. 2009; Van Raalte, interview, 10 June 2009.

198. Robert Dallos, "Solar Power Finds Place in Sun With Consumers," *Los Angeles Times*, 30 July 1976, A8; "A Battle against Time Is the Test at Optel," 22.

199. "Levitt Industries Merges into Optel in $18.1 Million Sale," *Wall Street Journal*, 3 July 1978, 4; Zanoni and Luce, interview, 24 Nov. 1998.

200. Castellano, *Liquid Gold*, 95; *Moody's OTC Industrial Manual: 1979* (New York: Moody's Investors Service, 1979), 1228; Zanoni and Luce, interview, 24 Nov. 1998.

201. Kiss, interview, 20 Sept. 2013.

202. Christensen, *Innovator's Dilemma*, 191–93.

203. Jill Lepore, "The Disruption Machine," *New Yorker*, 23 June 2014, 34.

204. E.g., "Electronic Watches for the Mass Market," 62; "Digital Watches: Bringing Watchmaking Back to the U.S.," *Business Week*, 27 Oct. 1975, 78; "The Great Digital Watch Shake-Out," *Business Week*, 2 May 1977, 78; Victor K. McElheny, "The Shakeout in Digital Watches," *New York Times*, 21 Sept. 1977, D1, D7.

205. Christensen, *Innovator's Dilemma*, 117–20.

206. Lepore, "Disruption Machine,"34–36.

207. George H. Heilmeier, "Liquid Crystal Displays: An Experiment in Interdisciplinary Research That Worked," *IEEE Transactions on Electron Devices* ED-23, no. 7 (July 1976): 784.

208. Ibid., 785.

Conclusion

1. Kenneth Bilby, *The General: David Sarnoff and the Rise of the Communications Industry* (New York: Harper and Row, 1986), chap. 14; Robert Sobel, *RCA* (New York: Stein and Day, 1986), chap. 14.

2. Barnaby J. Feder, "G.E. Makes a Gift of RCA Lab," *New York Times*, 6 Feb. 1987, D1–D2. SRI International was founded as the Stanford Research Institute in 1946. It became independent of the university in 1970 and adopted its current name in 1977.

3. Joel Brinkley, *Defining Vision: How Broadcasters Lured the Government into Inciting a Revolution in Television* (New York: Harcourt Brace, 1997); Thomas V. Lento, ed., *Inventing the Future: 60 Years of Innovation at Sarnoff* (Princeton, NJ: Sarnoff Corporation, 2006).

4. The full text of all three plaques can be found at "Milestones: List of IEEE Milestones," Engineering and Technology History Wiki, http://ethw.org/Milestones:List_of_IEEE_Milestones.

5. Ibid.

6. Joseph A. Castellano, *Liquid Gold: The Story of Liquid Crystal Displays and the Creation of an Industry* (Hackensack, NJ: World Scientific, 2005), 101.

7. Castellano, *Liquid Gold*, 132–34. Oh is first listed as a member of the DSRC liquid crystal group in RCA Laboratories, *Progress Report: Liquid Crystals, Jan.–Mar. 1970*, David Sarnoff Library Collection, Hagley Museum and Library, Wilmington, DE (DSL).

8. Castellano, *Liquid Gold*, 97–100.

9. Ibid., 111–12.

10. Castellano, *Liquid Gold*, 142–44; Arnold Thackray and Minor Myers Jr., *Arnold O. Beckman: One Hundred Years of Excellence* (Philadelphia: Chemical Heritage Foundation, 2000), 300–301.

11. Castellano, *Liquid Gold*, 112–18.

12. Castellano, *Liquid Gold*, 143–44; Thackray and Myers, *Arnold O. Beckman*, 300; Dietrich W. Grabis, "The LCD Market: The Origins and Underlying Reasons for the Growth of Liquid Crystal Display Technology in the USA—Focus: Electronic Watch and Calculator Applications" (DBA diss., Golden Gate University, San Francisco, 1980), 115, 161, 174.

13. Thackray and Myers, *Arnold O. Beckman*, 300.

14. Castellano, *Liquid Gold*, 144.

15. Ibid., 118–25.

16. Ross Knox Bassett, *To the Digital Age: Research Labs, Start-Up Companies and the Rise of MOS Technology* (Baltimore: Johns Hopkins University Press, 2002), 201–2, 374n95.

17. Quoted in Arnold Thackray, David C. Brock, and Rachel Jones, *Moore's Law: The Life of Gordon Moore, Silicon Valley's Quiet Revolutionary* (New York: Basic Books, 2015),335.

18. "Digital Watches: Bringing Watchmaking Back to the U.S.," *Business Week*, 27 Oct. 1975, 80; Victor K. McElheny, "The Shakeout in Digital Watches," *New York Times*, 21 Sept. 1977, D1.

19. Castellano, *Liquid Gold*, 107; Lucien F. Trueb, Günther Ramm, and Peter Wenzig, *Electrifying the Wristwatch* (Atglen, PA: Schiffer, 2013), 285–87.

20. "The $20 Digital Watch Arrives a Year Early," *Business Week*, 26 Jan. 1976, 27–28; Kathleen McDermott, *Timex: A Company and Its Community, 1854–1998* (New York: Timex Corporation, 1998), 193.

21. "The LCD Digital Makes a Comeback," *Business Week*, 19 Apr. 1976, 40.

22. Alan Sussman, interview by author, 19 Jan. 2011, Princeton, NJ; Robert Ricci, e-mail message to author, 7 Jan. 2011. Timex assigned Mr. Ricci as Industrial Relations Manager at the Franklin Township plant. He confirmed that most of the other employees in the facility had previously worked at RCA.

23. Grabis, "LCD Market," 51.

24. Ibid., 112.

25. "The Great Digital Watch Shake-Out," *Business Week*, 2 May 1977, 78.

26. Trueb, Ramm, and Wenzig, *Electrifying the Wristwatch*, 297–300; "Layoff Set by Texas Instruments," *New York Times*, 30 May 1981, 29.

27. Castellano, *Liquid Gold*, 134; McDermott, *Timex*, 207; Trueb, Ramm, and Wenzig, *Electrifying the Wristwatch*, 300–301.

28. For more on Heilmeier's time at DARPA, with a specific emphasis on computing, see George H. Heilmeier, interview by Arthur L. Norberg, 27 Mar. 1991, Livingston, NJ, Charles Babbage Institute, retrieved from the University of Minnesota Digital Conservancy, http://hdl.handle.net/11299/107352.

29. George H. Heilmeier, "Displays: A Pentagon Perspective," *IEEE Transactions on Electron Devices* ED-20, no. 11 (Nov. 1973): 923.

30. Bob Johnstone, *We Were Burning: Japanese Entrepreneurs and the Forging of the Electronic Age* (New York: Basic Books, 1999), 104–5.

31. David Hounshell, "The Evolution of Industrial Research in the United States," in *Engines of Innovation: U.S. Industrial Research at the End of an Era*, ed. Richard S. Rosenbloom and William J. Spencer (Boston: Harvard Business School Press, 1996), 50–51.

32. Alfred D. Chandler Jr., *Inventing the Electronic Century: The Epic Story of the Consumer Electronics and Computer Industries* (New York: Free Press, 2001), 1–12, 34–36. See also Hyungsub Choi, "Technology Importation, Corporate Strategies, and the Rise of the Japanese Semiconductor Industry in the 1950s," *Comparative Technology Transfer and Society* 6, no. 2 (Aug. 2008): 103–26.

33. These numbers are derived from "Liquid Crystal Conference Attendance—Kent State University, August 16–20, 1965" in Liquid Crystal Institute Records, 1965–1998, Box 1, Folder 6, Kent State University Special Collections.

34. George W. Gray, "Reminiscences from a Life with Liquid Crystals," *Liquid Crystals* 24, no 1 (1998): 6.

35. G. W. Gray, *Molecular Structure and the Properties of Liquid Crystals* (New York: Academic Press, 1962).

36. Cyril Hilsum, "The Anatomy of a Discovery: Biphenyl Liquid Crystals," in *Technology of Chemicals and Materials for Electronics*, ed. E. R. Howells (London: Society of Chemical Industry, 1984), 44.

37. Hilsum, "Anatomy of a Discovery," 46.

38. Gray, "Reminiscences," 6.

39. Hilsum, "Anatomy of a Discovery," 47.

40. G. W. Gray, "Development of Liquid Crystal Materials for Information Technology," in *Milestones in 150 Years of the Chemical Industry*, ed. P. J. T. Morris, W. A. Campbell, and H. L. Roberts (Cambridge: Royal Society of Chemistry, 1991), 292.

41. Hilsum, "Anatomy of a Discovery," 47–51.

42. As Castellano notes, the Darmstadt-based business that purchased BDH (E. Merck) should not be confused with the New Jersey–based Merck & Company, though the two were part of the same firm before World War I. (129)

43. Manuela Bremshey-Wilhelm, *Coincidence and Courage—1968–2008: 40 Years of Liquid Crystals Research at Merck* (Frankfurt am Main: Trademark, 2008), 41–45.

44. Hilsum, *Anatomy of a Discovery*, 52.

45. Ibid., 53–54.

46. David Dunmur and Tim Sluckin, *Soap, Science and Flat-Screen TVs: A History of Liquid Crystals* (New York: Oxford University Press, 2011 [2014]), 224–25.

47. Choi, "Technology Importation," 107–9, 115–18, 121–23.

48. Johnstone, *We Were Burning*, 10–12.

49. M. C. Steele, "Laboratories RCA, Inc. . . . RCA Research in Japan," *RCA Engineer* 9, no. 1 (June–July 1963): 8–9; Bernard Hershenov, "RCA Research Laboratories, Inc. (Tokyo)," *RCA Engineer* 20, no. 3 (Oct.–Nov. 1974): 24–32.

50. Hershenov, "RCA Research Laboratories," 24, 29–31.

51. Steele, "Laboratories RCA," 9.

52. Johnstone, *We Were Burning*, 89.

53. Hirohisa Kawamoto, "The History of Liquid-Crystal Displays," *Proceedings of the IEEE* 90, no. 4 (Apr. 2002): 468. Lucian Barton recorded that "TV people from Japan" came to visit the DSRC on 28 Aug. 1968 in Notebook 37998, p. 69, DSL.

54. Quoted in Hedrick Smith, *Rethinking America* (New York: Random House, 1995), 19.

55. Johnstone, *We Were Burning*, 89–91; Kawamoto, "History of Liquid-Crystal Displays," 483.

56. Kawamoto, "History of Liquid-Crystal Displays," 483.

57. Kawamoto, "History of Liquid-Crystal Displays," 483–86; Martin Schadt, "Milestone in the History of Field-Effect Liquid Crystal Displays and Materials," *Japanese Journal of Applied Physics* 48 (2009), doi:org/10.1143/JJAP.48.03B001, 4.

58. Johnstone, *We Were Burning*, 37–60.

59. Johnstone, *We Were Burning*, 91–94; Kawamoto, "History of Liquid-Crystal Displays," 468.

60. Tadashi Sasaki, interview by William Aspray, 25 May 1994, IEEE History Center, Hoboken, NJ, http://ethw.org/Oral-History:Tadashi_Sasaki.

61. See chap. 5.

62. Smith, *Rethinking America*, 22. The new name derived from the Ever-Sharp mechanical pencils that company founder (and namesake) Hayakawa Tokuji invented in 1915. After the company entered the electronics market, it started selling radios, TVs, and calculators using the Sharp brand. For more, see Johnstone, *We Were Burning*, 32–42.

63. Kawamoto, "History of Liquid-Crystal Displays," 469. See also Claudia

Flavell-While, "Chemical Engineers Who Changed the World [Tomio Wada]," *Chemical Engineer*, no. 843 (Sept. 2011): 54–55.

64. Kawamoto, "History of Liquid-Crystal Displays," 468–70.

65. Kawamoto, "History of Liquid-Crystal Displays," 470–71, 481.

66. Johnstone, *We Were Burning*, 105.

67. Castellano, *Liquid Gold*, 192, 281–82; Johnstone, *We Were Burning*, 119–20; Kawamoto, "History of Liquid-Crystal Displays," 494.

68. T. Peter Brody, "The Birth and Early Childhood of Active Matrix: A Personal Memoir," *Journal of the Society for Information Display* 4, no. 3 (Oct. 1996): 113–27. Brody and his Westinghouse colleague Fang-Chen Luo received the Jun-ichi Nishizawa Medal in 2011, along with Lechner, for their contributions to LCD technology.

69. T. P. Brody, Fang Chen Luo, Zoltan P. Szepesi, and David H. Davies, "A 6 × 6 in 20-lpi Electroluminescent Display Panel," *IEEE Transactions on Electron Devices* ED-22, no. 9 (Sept. 1975): 739–48.

70. As noted in chap. 3, Lechner's team had considered the use of thin-film transistors to address its liquid crystal displays in the 1960s. Paul Weimer, the RCA physicist who invented thin-film transistors, also wrote a patent for a flat-panel display that used these devices to activate either electroluminescent or liquid crystal cells. See Paul K. Weimer, Integrated Display Panel Utilizing Field-Effect Transistors, US Patent 3,564,135, filed 12 Oct. 1967 and issued 16 Feb. 1971.

71. Johnstone, *We Were Burning*, 138–41.

72. Kawamoto, "History of Liquid-Crystal Displays," 468–69, 494–95; Johnstone, *We Were Burning*, 140–42.

73. Castellano, *Liquid Gold*, 242–47; Paul O'Donovan, "Goodbye, CRT," *IEEE Spectrum* 43, no. 11 (Nov. 2006): 38–43; Larry F. Weber, "History of the Plasma Display Panel," *IEEE Trans. on Plasma Science* 34, no. 2 (Apr. 2006): 268–78.

74. Johnstone, *We Were Burning*, 143.

75. Castellano, *Liquid Gold*, 224–25; Lawrence E. Tannas Jr., "Flat-Panel Displays Displace Large, Heavy, Power-Hungry CRTs," *IEEE Spectrum* 26, no. 9 (Sept. 1989): 35; Ken Werner, "The Flowering of Flat Displays," *IEEE Spectrum* 34, no. 5 (May 1997): 44.

76. Castellano, *Liquid Gold*, 258; O'Donovan, "Goodbye, CRT,"; "LCD TVs Pass CRTs," *This Week in Consumer Electronics*, 25 Feb. 2008, 16.

77. Leslie Helm, "Japanese Companies Dominated the Market for LCD Screens But U.S. Electronics Firms Belatedly Are Gearing Up for Flat Out Competition," *Los Angeles Times*, 23 Jan. 1994, http://articles.latimes.com/print/1994-01-23/business/fi-18126_1_flat-panel-display.

78. Paul Gagnon (senior manager of analysis and research, IHS Markit), telephone conversation with author, 2 Feb. 2017.

79. Castellano, *Liquid Gold*, 69–70; Robert H. Chen, *Liquid Crystal Displays: Fundamental Physics and Technology* (Hoboken: Wiley and Sons, 2011), 416–17.

80. Castellano, *Liquid Gold*, 24, 95; Louis Zanoni and Nunzio (Tony) Luce, interview by Maggie Dennis and Carlene Stephens, 24 Nov. 1998, Trenton, NJ.

81. Castellano, *Liquid Gold*, 154–56, 232; Chen, *Liquid Crystal Displays*, 417–20, 424–25.

82. Gagnon, telephone conversation, 2 Feb. 2017.

83. Nicholson Baker, "A Fourth State of Matter," *New Yorker*, 8 and 15 July 2013, 70.

84. My use of the term *knowledge community* to describe a network of engineering practitioners unified around a common set of technical problems is derived from Ann Johnson, *Hitting the Brakes: Engineering Design and the Production of Knowledge* (Durham, NC: Duke University Press, 2009).

BIBLIOGRAPHY

Abbreviations

DSL	David Sarnoff Library Collection, Hagley Museum and Library, Wilmington,DE
IEEE	Institute of Electrical and Electronics Engineers
KPA	Zoltan Kiss Personal Archive, Belle Mead, NJ
ZPA	Louis Zanoni Personal Archive, Ewing, NJ

I. Oral Histories

All interviews conducted by the author unless otherwise indicated.

Briggs, George

12 Oct. 2009, Princeton, NJ

Caplan, Sandor

10 June 2009, Princeton, NJ

Channin, Donald

17 Aug. 2009, Princeton, NJ

Goldsmith, Norman

14 Aug. 2009, Princeton, NJ

Goodman, Lawrence

21 Aug. 2009, Princeton, NJ

Heilmeier, George

ca. 1970, interview by Stewart Finley, *Men and Molecules*, episode 515, "Liquid Crystals: A Bright Promise"

27 Mar. 1991, interview by Arthur L. Norberg, Livingston, NJ. Charles Babbage Institute, University of Minnesota. http://hdl.handle.net/11299/107352

4 Dec. 1998, interview by Margaret Dennis and Carlene Stephens, Morristown, NJ. National Museum of American History, Washington, DC

Helfrich, Wolfgang
28 Sept. 2009, Berlin, Germany
Hofstein, Steven
28 Oct. 2011, interview by David Laws, Mountain View, CA. Computer History Museum. http://archive.computerhistory.org/resources/access/text/2012/08/102745974-05-01-acc.pdf
Kiss, Zoltan
5 Sept. 2013, Belle Mead, NJ, in partnership with the Center for Oral History at the Chemical Heritage Foundation, Philadelphia
20 Sept. 2013, Belle Mead, NJ, in partnership with the Center for Oral History at the Chemical Heritage Foundation, Philadelphia
Klein, Richard
15 June 2009, Clifton, NJ, in partnership with the Center for Oral History at the Chemical Heritage Foundation, Philadelphia
Kornstein, Edward
17 Sept. 2014 (telephone conversation)
Lechner, Bernard
25 May 2004, interview by Alexander Magoun, Princeton, NJ
9 Apr. 2009, Princeton, NJ
17 Apr. 2009, Princeton, NJ
24 Apr. 2009, Princeton, NJ
Leverenz, Humboldt W.
15 July 1975, interview by Mark Heyer, IEEE History Center, Hoboken, NJ, http://ethw.org/Oral-History:Humboldt_W._Leverenz
Marlowe, Frank
24 Aug. 2009, Princeton, NJ
Matthies, Dennis
19 Aug. 2009, Princeton, NJ
Rajchman, Jan
26 Oct. 1970, interview by Richard Mertz, Archives Center, National Museum of American History, Washington, DC, http://amhistory.si.edu/archives/AC0196_rajc701026.pdf
11 July 1975, interview by Mark Heyer and Al Pinsky, Princeton, NJ. IEEE History Center, Hoboken NJ, http://ethw.org/Oral-History:Jan_Rajchman
Sasaki, Tadashi
25 May 1994, interview by William Aspray, Tokyo, Japan. IEEE History Center, Hoboken, NJ, http://ethw.org/Oral-History:Tadashi_Sasaki
Sussman, Alan
19 Jan. 2011, Princeton, NJ

Taylor, George
14 Oct. 2009, Pennington, NJ
Van Raalte, John
3 June 2009, Princeton, NJ
10 June 2009, Princeton, NJ
Webster, William
21 Apr. 2010 (telephone conversation)
3 Feb. 2011 (telephone conversation)
Williams, Richard
8 Apr. 2009, Princeton, NJ
16 Apr. 2009, Princeton, NJ
30 Apr. 2009, Princeton, NJ
6 May 2009, Princeton, NJ
Zaininger, Karl
7 Aug. 2009, Princeton, NJ
11 Sept. 2009, Princeton, NJ
Zanoni, Louis
1 July 2009, Ewing, NJ
21 Aug. 2009, Ewing, NJ
Zanoni, Louis, and Nunzio (Tony) Luce
24 Nov. 1998, interview by Margaret Dennis and Carlene Stephens, Trenton, NJ. National Museum of American History, Washington, DC

II. Archival Documents

A. SELECTED LABORATORY NOTEBOOKS (DSL)

Barton, Lucian
Notebook 22987 (25 Oct. 1963 to 11 Mar. 1964)
Notebook 24188 (11 Mar. 1964 to 9 Feb. 1965)
Notebook 26876 (10 Feb. 1965 to 1 Sept. 1965)
Notebook 28732 (1 Sept. 1965 to 3 Mar. 1966)
Notebook 30811 (3 Mar. 1966 to 22 June 1966)
Notebook 31917 (22 June 1966 to 23 Nov. 1966)
Notebook 33413 (25 Nov. 1966 to 26 May 1967)
Notebook 35399 (26 May 1967 to 27 Oct. 1967)
Notebook 36483 (30 Oct. 1967 to 1 May 1968)
Notebook 37998 (2 May 1968 to 10 Oct. 1968)
Notebook 39677 (11 Oct. 1968 to 27 Feb. 1969)
Notebook 40441 (27 Feb. 1969 to 8 Sept. 1969)

Notebook 42129 (9 Sept. 1969 to 13 Feb. 1970)
Notebook 43173 (16 Feb. 1970 to 18 June 1970)

Caplan, Sandor
Notebook 25716 (20 Aug. 1964 to 28 Dec. 1964)

Castellano, Joseph
Notebook 28304 (14 June 1965 to 22 June 1966)
Notebook 31913 (1 July 1966 to 29 Apr. 1967)
Notebook 34643 (28 Apr. 1967 to 25 Oct. 1968)
Notebook 4646 (29 Apr. 1967 to 21 Feb. 1968)
Notebook 37647 (21 Feb. 1968 to 17 Oct. 1968)
Notebook 39691 (17 Oct. 1968 to 16 Apr. 1969)
Notebook 41600 (8 July 1969 to 21 Dec. 1970)
Notebook 45534 (22 Dec. 1970 to 28 June 1972)

Fatuzzo, Ennio
Notebook 15788 (5 Oct. 1961 to 20 Mar. 1962)

Friel, Ronald
Notebook 44437 (2 Dec. 1970 to 20 Jan. 1988)

Goldmacher, Joel
Notebook 21630 (10 June 1963 to 29 July 1964)
Notebook 25011 (10 Aug. 1964 to 9 Mar. 1965)
Notebook 27361 (11 Mar. 1965 to 14 Jan. 1966)
Notebook 30326 (20 Jan. 1966 to 21 Apr. 1967)
Notebook 34653 (5 May 1967 to 21 Feb. 1968)
Notebook 37646 (21 Feb. 1968 to 12 May 1969)

Heilmeier, George
Notebook 14445 (24 May 1961 to 6 Dec. 1961)
Notebook 15988 (15 Dec. 1961 to 7 Aug. 1962)
Notebook 18725 (10 Aug. 1962 to 11 Dec. 1962)
Notebook 19929 (14 Dec. 1962 to 23 Apr. 1963)
Notebook 20858 (23 Apr. 1963 to 23 Sept. 1963)
Notebook 22280 (23 Sept. 1963 to 29 Feb. 1964)
Notebook 24160 (2 Mar. 1964 to 28 July 1964)
Notebook 25463 (23 July 1964 to 5 Jan. 1965)
Notebook 26848 (5 Jan. 1965 to 7 May 1965)
Notebook 28088 (10 May 1965 to 28 Oct. 1965)
Notebook 29275 (29 Oct. 1965 to 21 Mar. 1966)
Notebook 30812 (1 Mar. 1966 to 30 Aug. 1966)
Notebook 32284 (31 Aug. 1966 to 15 Feb. 1967)
Notebook 34299 (15 Feb. 1967 to 29 June 1967)
Notebook 35430 (30 June 1967 to 2 Jan. 1968)

Notebook 37231 (3 Jan. 1968 to 26 Aug. 1968)
Notebook 39358 (11 Sept. 1968 to 11 July 1969)
Notebook 41619 (14 July 1969 to 6 Aug. 1970)

Hofstein, Steven
Notebook 31938 (5 Aug. 1966 to 25 Mar. 1968)

Kane, Jean
Notebook 26045 (1 Oct. 1964 to 28 Jan. 1965)
Notebook 26883 (1 Feb. 1965 to 18 June 1965)
Notebook 28090 (14 May 1965 to 22 June 1966)
Notebook 31729 (1 June 1966 to 10 Nov. 1966)
Notebook 33180 (22 Nov. 1966 to 4 Apr. 1967)

Larach, Simon
Notebook 982 (1 Oct. 1956 to 30 Mar. 1966)

Lechner, Bernard
Notebook 3775 (25 June 1957 to 4 Nov. 1957)
Notebook 4084 (6 Nov. 1957 to 3 Mar. 1959)
Notebook 8593 (30 July 1959 to 16 Sept. 1959)

Lohman, Robert
Notebook 32487 (Oct. 1966 to Dec. 1967)

Matthies, Dennis
Notebook 39857 (19 Nov. 1968 to 26 Feb. 1969)

Moles Warren
Notebook 26365 (4. Dec. 1964 to 18 Jan. 1966)

Oh, Chan Soo
Notebook 45427 (10 Nov. 1970 to 27 Apr. 1971)
Notebook 45899 (27 Jan. 1971 to 6 May 1971)
Notebook 46471 (6 May 1971 to 16 Nov. 1971)

Rajchman, Jan
File 334 (12 Nov. 1951 to 19 May 1953)
P-2519 (Aug. 1952 to Jan. 1953)
Research Records PDP-5 (Bound 22 Sept. 1953)
Research Records PDP-6 (Bound 22 Sept. 1953)
Notebook 41 (4 Nov. 1954 to 17 Aug. 1955)
Notebook 722 (23 Aug. 1955 to 10 Sept. 1956)
Notebook 1620 (10 Sept. 1956 to 21 Dec. 1959)

Sussman, Alan
Notebook 41395 (7 July 1969 to 4 Feb. 1971)
Notebook 46488 (9 Feb. 1971 to 1 Nov. 1971)

Taylor, George
Notebook 18717 (1 Aug. 1962 to 24 May 1963)

Notebook 21385 (24 May 1963 to 28 Sept. 1964)
Notebook 25779 (28 Sept. 1964 to 14 Dec. 1964)
Notebook 27204 (16 Dec. 1964 to 27 Oct. 1965)
Notebook 29288 (2 Nov. 1965 to 1 June 1966)
Notebook 34467 (1 June 1966 to 2 Apr. 1969)

Tults, Juri
Notebook 24825 (1 June 1964 to 15 Nov. 1965)
Notebook 29309 (15 Nov. 1965 to 4 May 1967)

Van Raalte, John
Notebook 25479 (12 Aug. 1964 to 23 Dec. 1964)
Notebook 26374 (15 Dec. 1964 to 25 May 1965)
Notebook 27689 (28 Apr. 1965 to 31 Dec. 1965)
Notebook 30300 (3 Jan. 1966 to 20 Sept. 1966)
Notebook 32280 (19 July 1966 to 28 Sept. 1966)
Notebook 32465 (14 Oct. 1966 to 2 May 1967)
Notebook 34669 (3 May 1967 to 3 Apr. 1969)
Notebook 37196 (28 Nov. 1967 to 10 Feb. 1970)
Notebook 40277 (2 Jan. 1969 to 16 Oct. 1969)
Notebook 43307 (12 Feb. 1970 to 29 Apr. 1970)
Notebook 44045 (30 Apr. 1970 to 12 Aug. 1970)
Notebook 44609 (12 Aug. 1970 to 25 Nov. 1970)
Notebook 45498 (1 Dec. 1970 to 6 Mar. 1971)
Notebook 46168 (15 Mar. 1971 to 31 Aug. 1971)
Notebook 47380 (31 Aug. 1971 to 14 Dec. 1971)
Notebook 47807 (14 Dec. 1971 to 21 Mar. 1972)
Notebook 48060 (21 Mar. 1972 to 1 Sept. 1972)
Notebook 49034 (1 Sept. 1972 to 16 Aug. 1973)
Notebook 50932 (21 Aug. 1973 to 14 Nov. 1974)
Notebook 52813 (21 Jan. 1975 to 10 Mar. 1981)

Williams, Richard
Notebook 15811 (10 Oct. 1961 to 24 May 1962)
Notebook 17672 (24 May 1962 to 5 Feb. 1963)
Notebook 20170 (5 Feb. 1963 to 8 Aug. 1963)
Notebook 17943 (19 Aug. 1963 to 11 Aug. 1964)
Notebook 25699 (21 Aug. 1964 to 12 Mar. 1965)
Notebook 27113 (15 Feb. 1965 to 18 May 1966)
Notebook 27372 (11 Mar. 1965 to 1 Feb. 1967)
Notebook 30814 (7 Mar. 1966 to 17 Mar. 1967)
Notebook 34286 (3 Feb. 1967 to 21 Oct. 1968)
Notebook 39655 (24 Sept. 1968 to 9 July 1971)

Zanoni, Louis

Notebook 16971 (9 Mar. 1962 to 11 May 1965)

Notebook 27707 (11 May 1965 to 18 Aug. 1966)

B. RCA TECHNICAL REPORTS AND ENGINEERING MEMORANDA (DSL)

Allwine, H. S., D. C. Darling, C. W. Henderson, S. Lasof, H. B. Law, and E. G. Ramberg. *An Experimental and Theoretical Investigation of a Flat Picture Tube*. PEM-1012C, 4 Oct. 1955.

Barton, L. A., and W. H. Moles. *Memory Characteristics of Photoconductor Materials*. PEM-2484, 6 Aug. 1964.

Castellano, Joseph A., Lawrence A. Goodman, and Ronald N. Friel. *Human Factor Analysis of Reflective Liquid Crystal Displays*. PRRL-72-TR-137, 30 Aug. 1972.

Heilmeier, George H., and Wolfgang Liebmann. *A Note on the Crystal Growth and Evaluation of Hexamine as an Optical Modulator*. PEM-2297, 9 May 1963.

Heilmeier, George H., Louis A. Zanoni, and Lucian A. Barton. *Dynamic Scattering . . . A New Electro-Optic Effect in Certain Classes of Nematic Liquid Crystals*. PTR-2366, 1 Nov. 1967.

Helfrich, W. *Alignment Inversion Walls in Nematic Liquid Crystals in the Presence of a Magnetic Field*. PTR-2567, 12 Sept. 1968.

———. *Capillary Flow of Cholesteric and Smectic Liquid Crystals*. PTR-2736, 15 May 1969.

———. *Capillary Viscometry of Cholesteric Liquid Crystals*. PEM-3187, 20 Oct. 1969.

———. *Conduction-Induced Alignment in Nematic Liquid Crystals: Basic Model and Stability Considerations*. PTR-2686, 26 Feb. 1969.

———. *Deformation of Cholesteric Liquid Crystals with Low Threshold Voltage*. PRRL-70-TR-164, 12 Aug. 1970.

———. *Electrohydrodynamic and Dielectric Instabilities of Cholesteric Liquid Crystals*. PRRL-70-TR-236, 4 Nov. 1970.

———. *A Molecular Theory of Flow Alignment of Nematic Liquid Crystals*. PTR-2473, 24 May 1968.

Kazan, B., and F. H. Nicoll. *An Electroluminescent Light-Amplifying Picture Panel*, PTR-458, 16 May 1955.

Lasof, Solomon, and David Epstein. *Color Television Projectors for the Home*. PTR-509, 25 Oct. 1955.

Pritchard, D. H. *Electro-Optic Light Valve TV Display*. PTR-2205, 4 Jan. 1967.

Rajchman, Jan A. *Mural Television*. PEM-1013C, 26 Oct. 1955.

Rajchman, Jan A., and A. W. Lo. *The Transfluxor*. PTR-473, 20 May 1955.

Schwartz, J. W. *Line Storage Gas Discharge Picture Reproducer: Practical Approach to Panel Television?* PEM-719, 28 June 1956.

Van Raalte, John. *SLV: A New TV Projection System*. PTR-2698, 27 Mar. 1969.

C. DAVID SARNOFF RESEARCH CENTER DOCUMENTS AND RCA ANNUAL REPORTS (DSL)

"Organization List, RCA Laboratories Division." Jan. 1951–Jan. 1959, Mar. 1966–Nov. 1968.

RCA Corporation. *Annual Report* [1951–1976]. New York: RCA, 1952–1977.

RCA Laboratories. *Progress Report: Liquid Crystals* [1965–1970].

———. *Research Report* [1951–1975].

———. *Staff Directory* [May 1960–Dec. 1976].

D. OTHER ARCHIVAL MATERIALS

"Agreement between Timex Corporation and RCA Corporation for Purchase and Sale of RCA Corporation's Liquid Crystal Display Operations." 31 Mar. 1976.

American Express Company. "Introducing the Quartz Segtronic." Mail-order brochure, ca. 1975. ZPA.

———. "Introducing the Ultimatic for men. And the Ultimatic for women." Mail-order brochure, ca. 1975. ZPA.

Bosomworth, Douglas, Lawrence A. Goldmuntz, Edward Kornstein, and Edwin Robbins. "Letter to the Stockholders of Quantel Corporation." 1 Dec. 1970. ZPA.

Castellano, Joseph A. *Liquid Crystal Research Program: 1972*. Undated. Joseph Castellano Collection. DSL.

———. *Liquid Crystal Research Program: 1972 Annual Report*. Jan. 1973. Joseph Castellano Collection, DSL.

———. *Liquid Crystal Research Program: Progress Report, Jan.–Feb. 1972*. Joseph Castellano Collection. DSL.

Castellano, J. A., et al. *Electronically Tuned Optical Filters*. Contract no. NAS-12-638. Apr. 1969. DSL.

———. *Electronically Tuned Optical Filters*. Contract no. NAS-12-638. Apr. 1970. DSL.

———. *Electronically Tuned Optical Filters*. NASA CR-112032. Jan. 1972. DSL.

———. *Field Effect Liquid Crystal*. RADC-TR-70-96. Defense Technical Information Center, Ft. Belvoir, VA. http://www.dtic.mil/dtic/tr/fulltext/u2/871971.pdf.

———. *Field-Effect Liquid Crystal: Final Report*. Jan. 1970. DSL.

———. *Liquid Crystal Systems for Electro-Optical Storage Effects*. AFML-TR-73-51. Dec. 1971. Defense Technical Information Center, Ft. Belvoir, VA. http://www.dtic.mil/dtic/tr/fulltext/u2/760173.pdf.

David Sarnoff Library. *Accession Records*. 24 Aug. 1962, DSL.

Elms, James. "Electronic Research Center Weekly Highlights Report" [Memorandum to NASA Headquarters]. 6 Jun. 1968, Washington National Records Center. "Chronological File 1968," Box 3, Acc. 255-71A-2309.

Engstrom, Elmer. "Horizons in Electronics—A Revolution in Materials." WESCON, San Francisco, 26 Aug. 1955. Elmer W. Engstrom Papers, Box 1, Vol. 8, Dept. of Rare Books and Special Collections, Princeton University Library.

Eulogies, David Sarnoff 1891–1971. David Lilienthal Papers, Box 498: 1971 Correspondence: Name Files (S–Z), RE: Sarnoff, David 1971, Public Policy Papers, Department of Rare Books and Special Collections, Princeton University Library.

"Fault Finding." *RCA Radiations* 21 (Winter 1974–1975): 2. DSL.

"General David Sarnoff, 1891–1971." *RCA Radiations* 18 (Winter 1971–1972): 3–7, DSL.

Harrison, S. E., et al. *Utilization of Organic Materials for Molecular Electronics-Interim Engineering Report No. 1*. Prepared under contract no. AF33(657)-11662. 15 Sept. 1963. DSL.

———. *Utilization of Organic Materials for Molecular Electronics-Interim Engineering Report No. 2*. Prepared under contract no. AF33(657)-11662. 15 Dec. 1963. DSL.

———. *Utilization of Organic Materials for Molecular Electronics-Interim Engineering Report No. 3*. Prepared under contract no. AF33(657)-11662. 15 Mar. 1964. DSL.

———. *Utilization of Organic Materials for Molecular Electronics-Interim Engineering Report No. 4*. Prepared under contract no. AF33(657)-11662. 15 Sept. 1964. DSL.

———. *Utilization of Organic Materials for Molecular Electronics: Technical Documentary Report No. AL TDR 64-178*. July 1964. DSL.

Heilmeier, George. "Press Conference Presentation." [28 May 1968]. Bernard Lechner Collection, DSL.

Hillier, James. "Draft of Remarks by Dr. Hillier at Liquid Crystal Press Conference, 28 May 1968." Bernard Lechner Collection. DSL.

Hinckley, R. H. "Long Range Plan Revision." 20 Aug. 1965. "Chronological File—August 1965," Box 10, Acc. 255-71A-2309. Washington National Records Center, Suitland, MD.

Hockings, E. F. *Glass and Organic Materials Research: Quarterly Progress Report, Dec. 1971–Feb. 1972*. Joseph Castellano Collection, DSL.

———. *Glass and Organic Materials Research: Quarterly Progress Report, Mar.–May 1972*. Joseph Castellano Collection, DSL.

———. *Glass and Organic Materials Research: Quarterly Progress Report, Sep.–Nov. 1972*. Joseph Castellano Collection, DSL.

Homer, Edward J. *Market Potential of Products Based on Liquid Crystal Dynamic Scattering*. New York: RCA Product and Marketing Planning, 30 Mar. 1970. DSL.

"International Liquid Crystal Conference." Program, Kent State University, 1965. Liquid Crystal Institute Records, 1965–1998, Box 1, Folder 6. Kent State University Special Collections.

Kilbon, Kenyon. "Pioneering in Electronics: A Short History of the Origins and Growth of RCA Laboratories, Radio Corporation of America, 1919 to 1964." Unpublished typescript, rev. Aug. 1964. http://www.davidsarnoff.org/kil.html.

Kiss, Zoltan J. Zoltan J. Kiss to Dr. Hans Widmer [Omega]. Letter dated 3 Mar. 1971. ZPA.

Lechner, B. J., et al. *Development of a Solid-State Matrix Display*. AFFDL-TR-67-71, July 1967. Defense Technical Information Center, Ft. Belvoir, VA.

———. *Preliminary Development of a Solid-State Matrix Display*. AFFDL-TR-66-5, Jan. 1967. Defense Technical Information Center, Ft. Belvoir, VA.

Lechner, Bernard. "'Suggested Research Program' on Preliminary Development of a Liquid Crystal Matrix Display." Ca. 1969. Bernard Lechner Collection. DSL.

Light Emitting Diode Displays. Technical Proposal SP-68-47. Somerville: RCA Electronic Components, 21 June 1968. DSL.

"Liquid Crystal Conference Attendance—Kent State University, August 16–20, 1965." Liquid Crystal Institute Records, 1965–1998, Box 1, Folder 6. Kent State University Special Collections.

Medina, Ramiro. Ramiro Medina to David J. Barnette [*sic*]. Letter dated 6 Nov. 1974. ZPA.

"Mirror, mirror . . ." *RCA Radiations* 16, no. 4 (July–Aug. 1969). DSL.

"Nicoll, Frederick H." DSRC Biographies Collection, Box 4, Folder 15. DSL.

Optel Corporation. *Annual Report 1973*. Princeton, NJ: Optel, 1974. ZPA.

———. *Interim Report for the Three Months Ended March 31, 1973*. ZPA.

———. *Optel: Pioneers in Electronic Digital Watch Technology*. Princeton, NJ: Optel, ca. 1975. KPA.

———. "Optel Reflicon Display Model D-10." 15 Mar. 1971. KPA.

———. "Optel Terminal Model T710." Ca. 1971. KPA.

Pinsky, Al. "Preliminary Plans for Press Conference to Announce RCA Development of Liquid Crystal Electron Optic [*sic*] Effect." Interdepartmental correspondence, 10 Apr. 1968. Bernard Lechner Collection, DSL.

Quantel Corporation. "Technical Proposal No. 700601-1: Development Program of a Liquid Crystal Watch Display for Bulova Watch Company." 1 June 1970. Refac Intl Ltd v. Hitachi Ltd, et al., 2:87-cv-06191-TJH, Accession 021-96-0107, Location 20015511, Box 194-202. National Archives at Riverside, Perris, CA.

Quinn, R. E. "Interdepartmental Correspondence: Liquid Crystal Display Report—January-March 1967." 1 Jun 1967. DSL.

Quinn, Robert L. "Zoltan's Last Supper." Cartoon dated 7 Nov. 1969, KPA.

RCA Corporation. "RCA: A Decade of Difference." Excerpt from original 16 mm. film. Computer History Museum. http://www.computerhistory.org/revolution/mainframe-computers/7/169/2273.

RCA Department of News and Information. "Fact Sheet: Liquid Crystal Displays." [28 May 1968]. ZPA.

RCA Electronic Data Processing Division. *RCA 501 Electronic Data Processing System*. [1958].http://archive.computerhistory.org/resources/text/RCA/RCA.501.1958.102646273.pdf.

RCA Liquid Crystals: Domestic Licensing. New York: [RCA], Sept. 1971. DSL.

RCA News. "RCA Announces Breakthrough in Liquid Crystal Field." Press release, 28 May 1968. DSL.

Sarnoff, David. *New Developments in Electronics*. New York: Department of Information, Radio Corporation of America, 1955. Correspondence. David E. Lilienthal Papers, Box 400, 1955 Name Files (O-Sp.), Public Policy Papers. Department of Rare Books and Special Collections, Princeton University Library, Princeton, NJ.

Spivak, J. F., to J. V. Regan. "Patent Coverage Related to Liquid Crystals." 27 June 1968. DSL.

Van Raalte, John. *RCA Laboratories' Research Program on Large-Area Flat TV Displays*. Nov. 1973. DSL.

"Where the Company Goes from Here." *Communicate: The Magazine of RCA*, July–Aug. 1973, 14–15. DSL.

Zoltan J. Kiss, RCA Laboratories, 1960–1969. [7 Nov. 1969]. KPA.

III. PATENTS

Caplan, Sandor. Liquid Crystal Day/Night Mirror. US Patent 3,614,210, filed 6 Nov. 1969 and issued 19 Oct. 1971.

Fergason, James. Display Devices Utilizing Liquid Crystal Light Modulation. US Patent 3,731,986, filed 22 Apr. 1971 and issued 8 May 1973.

Heilmeier, George H., and Louis A. Zanoni. Electro-Optical Device. US Patent 3,499,112, filed 31 Mar. 1967 and issued 3 Mar. 1970.

Helfrich, W., and M. Schadt. Optical Device, Swiss Patent 532,261, filed 4 Dec. 1970 and issued 15 Feb. 1972.

Hofstein, Steven R. Solid State Clock. US Patent 3,505,804, filed 23 Apr. 1968 and issued 14 Apr. 1970.

Klein, Richard I., Sandor Caplan, and Ralph T. Hansen. Liquid Crystal Display Device. US Patent 3,689,131, filed 29 June 1970 and issued 5 Sept. 1972.

Levin, Barnett, and Nyman Levin. Improvements in or Relating to Light Valves. UK Patent 441,724, filed 13 July 1934 and issued 13 Jan. 1936.

Rajchman, Jan A., and George R. Briggs. Electroluminescent Apparatus. US Patent 3,041,490, filed 31 May 1955 and issued 26 June 1962.

Rajchman, Jan Aleksander. Electrical Display Device. US Patent 2,928,894, filed 31 May 1955 and issued 15 Mar. 1960.

Weimer, Paul K. Integrated Display Panel Utilizing Field-Effect Transistors. US Patent 3,564,135, filed 12 Oct. 1967 and issued 16 Feb. 1971.

Williams, Richard. Electro-Optical Elements Utilizing an Organic Nematic Compound. US Patent 3,322,485, filed 9 Nov. 1962 and issued 30 May 1967.

Zanoni, Louis A. Color Advertising Display Employing Liquid Crystal. US Patent 3,576,364, filed 20 May 1969 and issued 27 Apr. 1971.

IV. BOOKS, ARTICLES, AND OTHER PUBLICATIONS

Abramson, Albert. *Zworykin, Pioneer of Television*. Urbana: University of Illinois Press, 1995.

Aftalion, Fred. *A History of the International Chemical Industry: From the "Early Days" to 2000*. 2nd ed. Philadelphia: Chemical Heritage Foundation Press, 2001.

Anderson, R. Joseph, and Orville R. Butler. *History of Physicists in Industry*. College Park, MD: American Institute of Physics, 2008. https://www.aip.org/history/pubs/HOPI_Final_report.pdf.

"Army Buys a Big Brain to Keep Track of Its Tank Parts." *New York Times*, 9 Dec. 1955, 41.

Asner, Glen Ross. "The Cold War and American Industrial Research." PhD diss., Carnegie Mellon University, 2006.

"Aspects of Broadcasting, Present and Future." *Radio Age* 13, no. 4 (Oct. 1954): 6–7, 32.

Baillod, Gil. "Horlogerie: Un Américain (Optel) à bout de souffle." *L'Impartial*, 27 Apr. 1976, 11.

———. "Optel demande un moratoire." *L'Impartial*, 25 June 1976, 11.

———. "SGT prêt à commercialiser une montre sans rouages." *L'Impartial*, 7 Mar. 1972, 1, 3.

Baker, Nicholson. "A Fourth State of Matter." *New Yorker*, 8 and 15 July 2013, 64–73.

Bassett, Ross Knox. *To the Digital Age: Research Labs, Start-Up Companies, and the Rise of MOS Technology*. Baltimore: Johns Hopkins University Press, 2002.

"A Battle against Time Is the Test at Optel." *Business Week*, 19 July 1976, 22–23.

Berlin, Isaiah. *The Hedgehog and the Fox: An Essay on Tolstoy's View of History*. London: Weidenfeld and Nicolson, 1953.

Bilby, Kenneth. *The General: David Sarnoff and the Rise of the Communications Industry*. New York: Harper and Row, 1986.

Blattner, D. J., and F. Sterzer. "Modulators and Demodulators for Laser Systems." *RCA Engineer* 8, no. 5 (Feb.–Mar. 1963): 16–19.

"Boost for Color TV." *Time*, 10 Nov. 1958, 76.

Bremshey-Wilhelm, Manuela. *Coincidence and Courage—1968–2008: 40 Years of Liquid Crystals Research at Merck*. Frankfurt am Main: Trademark, 2008.

Brinkley, Joel. *Defining Vision: How Broadcasters Lured the Government into Inciting a Revolution in Television*. New York: Harcourt Brace, 1997.

Brock, David, and Christophe Lécuyer. "Digital Foundations: The Making of Silicon-Gate Manufacturing Technology." *Technology and Culture* 53, no. 3 (July 2012): 561–97.

Brody, T. P., Fang Chen Luo, Zoltan P. Szepesi, and David H. Davies. "A 6 × 6 in 20-lpi Electroluminescent Display Panel." *IEEE Transactions on Electron Devices* ED-22, no. 9 (Sept. 1975): 739–48.

Brody, T. Peter. "The Birth and Early Childhood of Active Matrix: A Personal Memoir." *Journal of the Society for Information Display* 4, no. 3 (Oct. 1996): 113–27.

Brown, George H. *And Part of Which I Was: Recollections of a Research Engineer*. Princeton, NJ: Angus Cupar, 1982.

"Built-in Entertainment." *Electronic Age* 19, no. 2 (Spring 1960): 10.

"Bulova to Introduce Quartz Crystal Watch That Retails for $395." *Wall Street Journal*, 6 Dec. 1971, 17.

Burck, Charles G. "Optel's (Mis)adventures in Liquid Crystals." *Fortune*, Oct. 1973, 192–96, 201–2, 206.

Burns, John L. "Forty Years of Electronic Progress." *Electronic Age* 18, no. 3 (Summer 1959): 2–3.

"Burns Quits R.C.A. Presidency: Elmer Engstrom Is Elected." *New York Times*, 2 Dec. 1961, 28, 31.

Burns, Russell W. *The Struggle for Unity: Colour Television, the Formative Years*. London: Institution of Engineering and Technology, 2008.

"Calculator/Watch Fits on the Wrist." *Electronics*, 1 May 1975, 40–42.

Calhoun, Amy. "George Heilmeier." *Penn Engineering* (Fall 2009): 10–11.

Carlson, W. Bernard. *Tesla: Inventor of the Electrical Age*. Princeton, NJ: Princeton University Press, 2013.

Castellano, Joseph A. "Liquid Crystals for Electro-Optical Application." *RCA Review* 33, no. 1 (Mar. 1972): 296–310.

———. *Liquid Gold: The Story of Liquid Crystal Displays and the Creation of an Industry*. Hackensack, NJ: World Scientific, 2005.

Chambet, Claude-Pierre. "Coup de maître! La Société des Garde-Temps va commercialiser, en première mondiale, une montre à quartz à affichage digital par cristaux liquides." *FAN-L'Express*, 7 Mar. 1972, 3.

———. "La petite histoire des cristaux liquides." *FAN-L'Express*, 2–3 Feb. 1974, 7.

———. "Si Optel ne paie pas ce qu'il doit à un industriel jurassien son stand risque d'être fermé ce matin à la Foire de Bâle!" *FAN-L'Express*, 28 Apr. 1976, 11.

———. "Un nom: Ditronic mais ni roue ni aiguille . . ." *FAN-L'Express*, 16 Mar. 1972, 3.

Chandler, Alfred D., Jr. *Inventing the Electronic Century: The Epic Story of the Consumer Electronics and Computer Industries*. New York: Free Press, 2001.

"Chasing the Rainbow." *Time*, 30 June 1958, 36, 39.

Chen, Robert H. *Liquid Crystal Displays: Fundamental Physics and Technology*. Hoboken: Wiley and Sons, 2011.

Choi, Hyungsub. "The Boundaries of Research: Making Transistors at RCA, 1948–1960." *Technology and Culture* 48, no. 4 (Oct. 2007): 758–82.

———. "Technology Importation, Corporate Strategies, and the Rise of the Japanese Semiconductor Industry in the 1950s." *Comparative Technology Transfer and Society* 6, no. 2 (Aug. 2008): 103–26.

Choi, Hyungsub, and Cyrus C. M. Mody. "The Long History of Molecular Electronics: Microelectronics Origins of Nanotechnology." *Social Studies of Science* 39, no. 1 (Feb. 2009): 11–50.

Christensen, Clayton M. *The Innovator's Dilemma*. New York: Harper Business, 2011.

Clarke, Sally H., Naomi R. Lamoreaux, and Steven W. Usselman, eds. *The Challenge of Remaining Innovative: Insights from Twentieth Century American Business*. Palo Alto, CA: Stanford Business Books, 2009.

"Color Television on Theatre Size Screen." *Radio Age* 11, no. 1 (Oct. 1951): 3–4, 31.

"Contributors." *Proceedings of the IRE* 45, no. 10 (Oct. 1957): 1422–23.

Cowie, Jefferson. *Capital Moves: RCA's Seventy-Year Quest for Cheap Labor*. Ithaca, NY: Cornell University Press, 1999.

Credelle, T. L. "Large-Screen Flat-Panel Television: A New Approach." *RCA Engineer* 26, no. 7 (July–Aug. 1981): 75–81.

Crisler, B. R. "New York Sees Color Television on Theater-Size Screen." *Christian Science Monitor*, 17 Oct. 1951, 7.

Curtis, Philip J. *The Fall of the U.S. Consumer Electronics Industry*. Westport, CT: Quorum Books, 1994.

Dallos, Robert. "Solar Power Finds Place in Sun with Consumers." *Los Angeles Times*, 30 July 1976, A6, A8.

"David Sarnoff of RCA Is Dead: Visionary Broadcast Pioneer." *New York Times*, 13 Dec. 1971, 1, 43.

Davis, Watson. "Electric Light without Bulbs." *Science News Letter* 59, no. 25 (23 June 1951): 394–95.

Dear, Peter, ed. *The Literary Structure of Scientific Argument: Historical Studies*. Philadelphia: University of Pennsylvania Press, 1991.

Dedication of Plaque to David Sarnoff: Commemorating His 45 Years of Service in Radio and Naming the Princeton Laboratories of RCA as the David Sarnoff Research Center. Princeton, NJ: RCA Laboratories Division, 1951.

"Digital Watches: Bringing Watchmaking Back to the U.S." *Business Week*, 27 Oct. 1975, 78, 80–81, 83, 86, 91–92.

Dinner in Honor of David Sarnoff (Chairman of the Board, Radio Corporation of America): Commemorating His Fifty Years of Service to Radio, Television and Electronics. New York: Radio Corporation of America, 1956.

Dorfman, Dan. "Heard on the Street." *Wall Street Journal*, 5 Oct. 1971, 47.

"Dr. David Kleitman," *San Francisco Chronicle* (25 Feb. 2004). http://www.sfgate.com/news/article/KLEITMAN-Dr-David-2818193.php.

Dreher, Carl. *Sarnoff: An American Success*. New York: Quadrangle/New York Times Book Co., 1977.

"Dr. Elmer W. Engstrom Elected President of RCA." *RCA Engineer* 7, no. 4 (Dec. 1961–Jan. 1962): center spread (between pp. 30 and 31).

Duncan, R. C., Jr., Z. J. Kiss, and H. R. Lewis. "Sun-Pumped Continuous Laser." *RCA Engineer* 8, no. 5 (Feb.–Mar. 1963): 62.

Dunmur, David, and Tim Sluckin. *Soap, Science, and Flat-Screen TVs: A History of Liquid Crystals*. New York: Oxford University Press, 2011. Reprint 2014.

Editors of *Fortune*. *The Fabulous Future: America in 1980*. New York: E. P Dutton, 1956.

"E. J. Nally, 94, Dies: First R.C.A. Head." *New York Times*, 23 Sept. 1953, 31.

"Electrical Voltages Derived by RCA from Radioactive Materials." *Wall Street Journal*, 28 Sept. 1951, 16.

"Electronic Watches for the Mass Market." *Business Week*, 22 Apr. 1972, 62, 66.

"En attendant qu'Optel trouve de l'argent, ses créanciers suisses lui accordent un sursis jusqu'au 6 juin . . ." *FAN-L'Express*, 8–9 May 1976, 9.

"Encore une 'Digitale' d'Optel . . ." *FAN-L'Express*, 23 Apr. 1972, 7.

"Engineering News and Highlights." *RCA Engineer* 14, no. 3 (Oct.–Nov. 1968): 94–96.

"Engineering News and Highlights." *RCA Engineer* 14, no. 4 (Dec. 1968–Jan. 1969): 93–96.

"Engineering News and Highlights." *RCA Engineer* 15, no. 1 (June–July 1969): 94–96.

"Engineering News and Highlights." *RCA Engineer* 20, no. 2 (Aug.–Sept. 1974): 94–96.

Engstrom, Elmer W. "A History of Radio Corporation of America, Part II: The Years 1938–1958." *RCA Engineer* 4, no. 1 (June–July 1958): 28–34.

———. "A History of Radio Corporation of America, Part III: The Years 1958–1962." *RCA Engineer* 9, no. 1 (June–July 1963): 2–6.

"Engstrom's Selection as RCA's President Termed Move to Boost Lagging Earnings." *Wall Street Journal*, 4 Dec. 1961, 8.

"Faded Rainbow." *Time*, 22 Oct. 1956, 96–97.

Fairley, Peter. "Solar-Cell Squabble," *IEEE Spectrum* 45, no. 4 (Apr. 2008): 36–40.

Farina, P. L. "Digital Readouts." *RCA Engineer* 17, no. 6 (Apr.–May 1972): 12–16.

Feder, Barnaby J. "G.E. Makes Gift of RCA Lab." *New York Times*, 6 Feb. 1987, D1–D2.

Fergason, James L. "Liquid Crystals." *Scientific American* 211, no. 2 (Aug. 1964): 76–85.

Fink, Donald G. *Television Engineering*. 2nd ed. New York: McGraw-Hill, 1952.

Fisher, David, and Marshall Jon Fisher. *Tube: The Invention of Television*. Washington, DC: Counterpoint, 1996.

Fisher, Franklin M., James W. McKie, and Richard B. Mancke. *IBM and the U.S. Data Processing Industry: An Economic History*. New York: Praeger, 1983.

Flamm, Kenneth. *Creating the Computer: Government, Industry, and High Technology*. Washington, DC: Brookings Institution, 1988.

Flavell-While, Claudia. "Chemical Engineers Who Changed the World [Tomio Wada]." *Chemical Engineer*, no. 843 (Sept. 2011): 54–55.

Forrest, Stephen, Paul Burrows, and Mark Thompson. "The Dawn of Organic Electronics." *IEEE Spectrum* 37, no. 8 (Aug. 2000): 29–34.

Freedman, N. S. "Superconductive Materials and Magnets: A Review of Progress at RCA." *RCA Engineer* 12, no. 4 (Dec. 1966–Jan. 1967): 7–10.

Friedel, Georges. "Les états mésomorphes de la matière." *Annales des Physique* 18 (1922): 273–474. Translated as "The Mesomorphic States of Matter." In *Crystals That Flow: Classic Papers from the History of Liquid Crystals*, edited by Timothy J. Sluckin, David A Dunmur, and Horst Stegemeyer, 162–211. New York: Taylor and Francis, 2004.

Gardner, W. David. "Curtain Act at RCA." *Datamation* 18 (19 Mar. 1972): 34–41.

"The General." *Time*, 23 July 1951, 74–80.

"General David Sarnoff, Miracle Man of the Electronic Age." *Advertiser*, Oct. 1951, 13, 42.

Gertner, Jon. *The Idea Factory: Bell Labs and the Great Age of American Innovation*. New York: Penguin Press, 2012.

Godin, Benoit. "The Linear Model of Innovation: The Historical Construction of an Analytical Framework." *Science, Technology & Human Values* 31, no. 6 (Nov. 2006): 639–67.

Gold, Martin. "ILC's Heller Heads Optel: Kiss Stays as Chairman." *Electronic News*, 9 Feb. 1976, 46.

———. "LCDs, Semicons Involved in Timex Deal with RCA." *Electronic News*, 16 Feb. 1976, 1, 63.

———. "Lewis Appointed Optel President." *Electronic News*, 8 Apr. 1974, 1, 58.

Goldmacher, Joel, and Lucian A. Barton. "Liquid Crystals. I. Fluorinated Anils." *Journal of Organic Chemistry* 32, no. 2 (Feb. 1967): 476–77.

Golinski, Jan. *Making Natural Knowledge: Constructivism and the History of Science*. 2nd ed. Chicago: University of Chicago Press, 2005.

Goodman, L. A. "Passive Liquid Displays: Liquid Crystals, Electrophoretics, and Electrochromics." *IEEE Transactions on Consumer Electronics* CE-21, no. 3 (Aug. 1975): 247–59.

Grabis, Dietrich W. "The LCD Market: The Origins and Underlying Reasons for the Growth of Liquid Crystal Display Technology in the USA—Focus: Electronic Watch and Calculator Applications." DBA diss., Golden Gate University, San Francisco, 1980.

Graham, Margaret B. W. *The Business of Research: RCA and the VideoDisc*. New York: Cambridge University Press, 1986.

Gray, G. W. "Development of Liquid Crystal Materials for Information Technology." In *Milestones in 150 Years of the Chemical Industry*, edited by P. J. T. Morris, W. A. Campbell, and H. L. Roberts, 288–303. Cambridge: Royal Society of Chemistry, 1991.

———. *Molecular Structure and the Properties of Liquid Crystals*. New York: Academic Press, 1962.

———. "Reminiscences from a Life with Liquid Crystals." *Liquid Crystals* 24, no. 1 (1998): 5–13.

"The Great Digital Watch Shake-Out." *Business Week*, 2 May 1977, 78, 80.

Guimaraes, Amilcar. "The Beginning of the Liquid Crystal Watch." *National Association of Watch and Clock Collectors Bulletin* 321 (Aug. 1999): 479–83.

Heilmeier, G. H. "The Dielectric and Electrooptical Properties of a Molecular Crystal: Hexamine." *Applied Optics* 3, no. 11 (Nov. 1964): 1281–87.

———. "Liquid Crystals: The First Electronic Method for Controlling the Reflection of Light." *RCA Engineer* 15, no. 1 (June–July 1969): 14–18.

Heilmeier, G. H., and L. A. Zanoni. "Guest-Host Interactions in Nematic Liquid Crystals: A New Electro-Optic Effect." *Appied Physics Letters* 13, no. 3 (1 Aug. 1968): 91–92.

Heilmeier, G. H., L. A. Zanoni, and L. A. Barton. "Dynamic Scattering in Nematic Liquid Crystals." *Applied Physics Letters* 13, no. 1 (1 July 1968): 46–47.

Heilmeier, George H. "An Analysis of Parametric Amplification in Periodically Loaded Transmission Lines." MS thesis, Princeton University, 1960.

———. "Displays: A Pentagon Perspective." *IEEE Transactions on Electron Devices* ED-20, no. 11 (Nov. 1973): 921–25.

———. "Liquid Crystal Displays: An Experiment in Interdisciplinary Research That Worked." *IEEE Transactions on Electron Devices* ED-23, no. 7. (July 1976): 780–85.

———. "Semiconduction in Phthalocyanine: A Study in Organic Semiconduction." PhD diss., Princeton University, 1962.

———. "Transient Behavior of Domains in Liquid Crystals." *Journal of Chemical Physics* 44, no. 2 (15 Jan. 1966): 644–47.

Heilmeier, George H., and Joel E. Goldmacher. "Electric-Field-Induced Cholesteric-Nematic Phase Change in Liquid Crystals." *Journal of Chemical Physics* 51, no. 3 (Aug. 1969): 1258–60.

Heilmeier, George H., Louis A. Zanoni, and Lucian A. Barton. "Dynamic Scattering: A New Electrooptic Effect in Certain Classes of Nematic Liquid Crystals." *Proceedings of the IEEE* 56, no. 7 (July 1968): 1162–71.

Helm, Leslie. "Japanese Companies Dominated the Market for LCD Screens But U.S. Electronics Firms Belatedly Are Gearing Up for Flat Out Competition." *Los Angeles Times*, 23 Jan. 1994. http://articles.latimes.com/print/1994-01-23/business/fi-18126_1_flat-panel-display.

Hershenov, Bernard. "RCA Research Laboratories, Inc. (Tokyo)." *RCA Engineer* 20, no. 3 (Oct.–Nov. 1974): 24–32.

Hershey, Robert D., Jr. "Solar Power Race at Jersey Concern." *New York Times*, 2 July 1982, D1, D3.

Hilsum, Cyril. "The Anatomy of a Discovery: Biphenyl Liquid Crystals." In *Technology of Chemicals and Materials for Electronics*, edited by E. R. Howells, 43–58. London: Society of Chemical Industry, 1984.

Holmes, Frederic L., Jürgen Renn, and Hans-Jörg Rheinberger, eds. *Reworking the Bench: Research Notebooks in the History of Science*. Boston: Kluwer Academic, 2003.

Hornbeck, Larry J. "From Cathode Rays to Digital Micromirrors: A History of Electronic Projection Display Technology." *TI Technical Journal* (July–Sept. 1998): 7–46.

Hounshell, David. "The Evolution of Industrial Research in the United States." In *Engines of Innovation: U.S. Industrial Research at the End of an Era*, edited by Richard S. Rosenbloom and William J. Spencer, 13–85. Boston: Harvard Business School Press, 1996.

Hounshell, David, and John Kenly Smith Jr. *Science and Corporate Strategy: Du Pont R&D, 1902–1980*. New York: Cambridge University Press, 1988.

Howe, Tom. "Jay Brandinger." *Memories of VideoDisc: Who's Who in VideoDisc*, CEDmagic.com, http://www.cedmagic.com/mem/whos-who/brandinger-jay.html.

Howell, Jane M., and Christopher A. Higgins. "Champions of Change: Identifying, Understanding, and Supporting Champions of Technological Innovations." *Organizational Dynamics* 19, no. 1 (Summer 1990): 40–55.

Howells, E. R., ed. *Technology of Chemicals and Materials for Electronics*. London: Society of Chemical Industry, 1984.

"How to Install Mural TV." *Popular Science-Special Home Improvement Supplement* 175, no. 3 (Sept. 1959): 174–77.

Hughes, Thomas P. *American Genesis: A Century of Invention and Technological Enthusiasm, 1870–1970*. Chicago: University of Chicago Press, 2004. First published 1989 by Viking.

———. *Networks of Power: Electrification in Western Society, 1880–1930*. Baltimore: Johns Hopkins University Press, 1983.

"Introducing Your Engineering Leadership." *RCA Engineer* 1, no. 1 (June–July 1955): 25–27.

"In TV Future." *Washington Post*, 29 May 1968, D9.

"Inventions Wanted." *Time*, 8 Oct. 1951, 54.

Johnson, Ann. *Hitting the Brakes: Engineering Design and the Production of Knowledge*. Durham, NC: Duke University Press, 2009.

Johnston, Laurie. "700 at Sarnoff Service in Emanu-El." *New York Times*, 16 Dec. 1971, 38.

Johnstone, Bob. *Brilliant! Shuji Nakamura and the Revolution in Lighting Technology*. Amherst, NY: Prometheus Books, 2007.

———. *We Were Burning: Japanese Entrepreneurs and the Forging of the Electronic Age*. New York: Basic Books, 1999.

Jones, Stacy V. "New Liquid Crystal Is Devised." *New York Times*, 23 Feb. 1974, 39, 43.

———. "Testing System Is Found for Integrated Circuits." *New York Times*, 24 Jan. 1976, 37, 39.

Josephs, Jess. "A Review of Panel-Type Display Devices." *Proceedings of the IRE* 48, no. 8 (Aug. 1960): 1380–95.

K[aempffert]., W[aldemar]. "Bright Future for Electronics." *New York Times*, 24 Oct. 1954, E9.

Kawamoto, Hirohisa. "The History of Liquid-Crystal Displays." *Proceedings of the IEEE* 90, no. 4 (Apr. 2002): 460–500.

Kazan, B. "Electroluminescent Displays." *Proceedings of the SID* 17, no. 1 (1976): 23–29.

———. "A Solid-State Amplifying Fluoroscope Screen." *RCA Review* 19, no. 1 (Mar. 1958): 19–34.

Kazan, B. and F. H. Nicoll. "An Electroluminescent Light-Amplifying Picture Panel." *Proceedings of the IRE* 43, no. 12 (Dec. 1955): 1888–97.

Kenney, Harry C. "'Picture on Wall' Envisaged for Television." *Christian Science Monitor*, 23 Dec. 1954, 3.

Kihn, Harry. "Integrated Electronics . . . A Survey." *RCA Engineer* 7, no. 6 (Apr.–May 1962): 14–19.

King, Nicholas. "Young Scientists of Today." *Electronic Age* 28, no. 3 (Summer 1969): 28–30.

"Kiss Resigns as Optel Chairman." *Electronic News*, 26 Apr. 1976, 26.

Kiss, Zoltan J. "Photochromics." *Physics Today* 23, no. 1 (Jan. 1970): 42–49.

Kline, Ronald. "Construing 'Technology' as 'Applied Science': Public Rhetoric of Scientists and Engineers in the United States, 1880–1945." *Isis* 86, no. 2 (June 1995): 194–221.

Knoblauch, Mary. "Mood Ring Monitors Your State of Mind." *Chicago Tribune*, 8 Oct. 1975, C1.

Knowles, Scott G., and Stuart W. Leslie. "'Industrial Versailles': Eero Saarinen's Corporate Campuses for GM, IBM and AT&T." *Isis* 92, no. 1 (Mar. 2001): 1–33.

Koden, Mitsuhiro. *OLED Displays and Lighting*. Hoboken, NJ: John Wiley and Sons, 2016.

Kovaleff, Theodore. *Business and Government during the Eisenhower Administration*. Athens: Ohio University Press, 1980.

Kramer, Barry. "Chemical Chameleons: Liquid Crystals, Once Merely Curiosities, May Become Boon to Industry, Consumers." *Wall Street Journal*, 29 Feb. 1972, 36.

Kyoto Prize. "2005: Advanced Technology: Electronics: George H. Heilmeier." KyotoPrize.org. http://www.kyotoprize.org/en/laureates/george_h_heilmeier/.

Law, H. B. "A Review of Some Flat Display Devices." *RCA Engineer* 7, no. 3 (Oct.–Nov. 1961): 20–23.

Lawrence, W. F., and A. G. F. Dingwall. "Fossil-Fuel Thermoelectric Generators." *RCA Engineer* 9, no. 3 (Oct.–Nov. 1963): 16–20.

"Layoff Set by Texas Instruments." *New York Times*, 30 May 1981, 29, 32.

"The LCD Digital Makes a Comeback." *Business Week*, 19 Apr. 1976, 40.

"LCD TVs Pass CRTs." *This Week in Consumer Electronics*, 25 Feb. 2008, 16.

Lechner, Bernard. "History Crystallized: A First-Person Account of the Development of Matrix-Addressed LCDs for Television at RCA in the 1960s." *Information Display* (Jan. 2008): 26–30.

Lechner, Bernard J., Frank J. Marlowe, Edward O. Nester, and Juri Tults. "Liquid Crystal Matrix Displays." *Proceedings of the IEEE* 59, no. 11 (Nov. 1971): 1566–79.

Lécuyer, Christophe. *Making Silicon Valley: Innovation and the Growth of High Tech, 1930–1970*. Cambridge, MA: MIT Press, 2006.

Lécuyer, Christophe, and David C. Brock. *Makers of the Microchip: A Documentary History of Fairchild Semiconductor*. Cambridge, MA: MIT Press, 2010.

———. "The Materiality of Microelectronics." *History and Technology* 22, no. 3 (Sept. 2006): 301–25.

Lécuyer, Christophe, and Takahiro Ueyama. "The Logics of Materials Innovation: The Case of Gallium Nitride and Blue Light Emitting Diodes." *Historical Studies in the Natural Sciences* 43, no. 3 (June 2013): 243–80.

Lento, Thomas V., ed. *Inventing the Future: 60 Years of Innovation at Sarnoff*. Princeton, NJ: Sarnoff Corporation, 2006.

Lepore, Jill. "The Disruption Machine." *New Yorker*, 23 June 2014, 30–36.

Leverenz, Humboldt W. *An Introduction to the Luminescence of Solids*. New York: John Wiley and Sons, 1950.

"Levitt Industries Merges into Optel in $18.1 Million Sale." *Wall Street Journal*, 3 July 1978, 4.

Lewis, Anthony. "R.C.A. Is Indicted in Criminal Case under Trust Law." *New York Times*, 22 Feb. 1958, 1, 33.

Lewis, H. R. "The Laser: An Introduction." *RCA Engineer* 8, no. 5 (Feb.–Mar. 1963): 6–7.

———. "Quantum Electronics Research at RCA Laboratories: An Introduction." *RCA Engineer* 12, no. 3 (Oct.–Nov. 1966): 24.

Lewis, Thomas S. W. *Empire of the Air: The Men Who Made Radio*. New York: HarperCollins, 1991.

"Lights from Panelescence." *Wall Street Journal*, 14 July 1953, 13.

Lockhart, R. K. "Development of a 1000-MC Computer." *RCA Engineer* 6, no. 4 (Dec. 1960–Jan. 1961): 11–13.

"L''octoscope' ouvre la voie à la montre 100% electronique." *FAN-L'Express*, 15 Apr. 1971, 3.

"Loss on Color TV to R.C.A. $6,900,000." *New York Times*, 26 Dec. 1956, 38.

Luce, Nunzio A. "C/MOS Digital Watch Features Liquid Crystal Display." *Electronics*, 10 Apr. 1972, 93–97.

Lyons, Eugene. *David Sarnoff: A Biography*. New York: Harper and Row, 1966.

"Making Displays Easier to Read." *Business Week*, 21 Nov. 1970, 49.

"Management's Renaissance Man." *Time*, 7 Sept. 1959, 72.

Mauguin, Charles. "Sur les cristaux liquides de Lehmann." *Bulletin de la Société française de Minéralogie* 34 (1911): 71–117. Translated as "On the Liquid Crystals of Lehmann." In *Crystals That Flow: Classic Papers from the History of Liquid Crystals*, edited by Timothy J. Sluckin et al., 100–121. New York: Taylor and Francis, 2004.

Mayer, C. G. "Laboratories RCA, Ltd." *Radio Age* 15, no. 1 (Jan. 1956): 20–21.

McClure, Donald, and Zoltan Kiss. "Survey of the Spectra of Divalent Rare-Earth Ions in Cubic Crystals." *Journal of Chemical Physics* 39, no. 12 (15 Dec. 1963): 3251–57.

McDermott, Kathleen. *Timex: A Company and Its Community, 1854–1998*. New York: Timex Corporation, 1998.

McElheny, Victor K. "The Shakeout in Digital Watches." *New York Times*, 21 Sept. 1977, D1, D7.

Metz, Cade. "Liquid Crystals-Display Genius No Match for Petty Politics." *Register*, 27 Apr. 2009, http://www.theregister.co.uk/2009/04/27/george_heilmeier/.

"Milestones: List of IEEE Milestones." Engineering and Technology History Wiki. http://ethw.org/Milestones:List_of_IEEE_Milestones.

Moody's OTC Industrial Manual. New York: Moody's Investors Service, 1972–1979.

Moore, Gordon E. "Cramming More Components onto Integrated Circuits." *Electronics* 38, no. 8 (19 Apr. 1965): 114–17.

Moore, Kelly. *Disrupting Science: Social Movements, American Scientists, and the Politics of the Military, 1945–1975*. Princeton, NJ: Princeton University Press, 2008.

Morris, P. J. T., W. A. Campbell, and H. L. Roberts, eds. *Milestones in 150 Years of the Chemical Industry*. Cambridge: Royal Society of Chemistry, 1991.

Mullins, Morell E., Steve W. J. Kozlowski, Neal Schmitt, and Ann W. Howell. "The Role of the Idea Champion in Innovation: The Case of the Internet in the Mid-1990s." *Computers in Human Behavior* 24, no. 2 (Mar. 2008): 451–67.

Murray, Lawrence A., Sandor Caplan, and Richard Klein. "Lighting Up in a Group." *Electronics* 41, no. 5 (4 Mar. 1968): 104–10.

Myers, Greg. "From Discovery to Invention: The Writing and Rewriting of Two Patents." *Social Studies of Science* 25, no. 1 (Feb. 1995): 57–105.

National Science Board. *Science and Engineering Indicators 2016*. NSB-2016-1. Arlington, VA: National Science Foundation, 2016.

Neal, Terri Fergason, and Marian Pierce. *Mr. Liquid Crystal: The Untold Story of How James L. Fergason Invented the Liquid Crystal Display and Helped Create the Digital World*. Los Angeles: New Insights Press, 2016.

"New Solid-State Division Formed." *RCA Engineer* 15, no. 6 (Apr.–May 1970): 91.

"New Thin-Screen Displays Utilize Liquid Crystal Effect." *IEEE Spectrum* 5, no. 7 (July 1968): 129.

O'Donovan, Paul. "Goodbye, CRT." *IEEE Spectrum* 43, no. 11 (Nov. 2006): 38–43.

"Optel Corp. Elects Six to Board, Completes Stock-Debt Exchange." *Wall Street Journal*, 24 Dec. 1975, 10.

"Optel Corp. Offering Sells Out." *Wall Street Journal*, 22 June 1972, 29.

"Optel Corp. Said It Obtained Bank Credit of $1 Million." *Wall Street Journal*, 28 Mar. 1974, 23.

"Optel Corp. to Report Loss for Quarter, Year." *Wall Street Journal*, 11 Jan. 1974, 4.

"Optel Gets $1.5 Million Loan." *Wall Street Journal*, 26 July 1974, 17.

"Optel Gets Mitsubishi Loan." *Electronic News*, 29 July 1974, 17.

"Optel Says Demands from Creditors Force It to File for Chapter 11." *Wall Street Journal*, 17 June 1976, 10.

"Optel Says It's Seeking an Infusion of Capital to Ensure Its Survival." *Wall Street Journal*, 16 Apr. 1976, 21.

"Optel to Develop Auto Displays." *Electronic News*, 15 Oct. 1973, 52.

"Optoelectronics." *Electronic News*, 9 Feb. 1976, 1, 50–51, 54.

Peter, R. W. "Materials Research." *RCA Engineer* 5, no. 4 (Dec. 1959–Jan. 1960): 4–9.

"Philco Suit Cites Electronics 'Plot.'" *New York Times*, 15 Jan. 1957, 37, 40.

Phillips, William, and Zoltan J. Kiss. "Photo Erasable Dark Trace Cathode-Ray Storage Tube." *Proceedings of the IEEE* 56, no. 11 (Nov. 1968): 2072–73.

Plumb, Robert K. "Electronic Device Can Duplicate Every Sound." *New York Times*, 1 Feb. 1955, 1, 35.

———. "R.C.A. Unveils TV Tape Player, Light Amplifier, Silent Air Cooler." *New York Times*, 2 Oct. 1956, 37.

Poulos, Nick. "New Watches Will Tell More Than Time." *Chicago Tribune*, 20 Aug. 1975, C7.

"Products of the Future . . . ." *Radio Age* 14, no. 3 (July 1955): 16–17.

Pugh, Emerson. *Building IBM: Shaping an Industry and Its Technology*. Cambridge, MA: MIT Press, 1995.

Pulvari, C. F. "The Transpolarizer: An Electrostatically Controlled Circuit Impedance with Stored Setting." *Proceedings of the IRE* 47, no. 6 (June 1959): 1117–23.

Radio Corporation of America. *Working in Research at RCA*. Princeton, NJ: RCA Laboratories Division, 1956.

Rajchman, J. A., G. R. Briggs, and A. W. Lo. "Transfluxor Controlled Electroluminescent Display Panels." *Proceedings of the IRE* 46, no. 11 (Nov. 1958): 1808–24.

Rajchman, J. A., and A. W. Lo. "The Transfluxor." *Proceedings of the IRE* 44, no. 3 (Mar. 1956): 321–32.

Rajchman, Jan. "RCA Computer Research . . . Some History, and a Review of Current Work." *RCA Engineer* 8, no. 6 (Apr.–May 1963): 6–13.

Ramberg, E. G. "Electron-Optical Properties of a Flat Television Picture Tube." *Proceedings of the IRE* 48, no. 12 (Dec. 1960): 1952–60.

Ranzal, Edward. "R.C.A. Yields in Trust Suit: Will Ease Patent Licensing." *New York Times*, 29 Oct. 1958, 1, 39.

"R.C.A. and I.B.M. Sign Computer Patents Pact." *New York Times*, 11 Sept. 1957, 45.

"RCA Building New Jersey Plant." *Wall Street Journal*, 21 Feb. 1974, 8.

"R.C.A. Claims Big TV Advance in Light Device." *Chicago Tribune*, 20 Dec. 1954, C7.

"RCA Develops a New Visual Display Means Using Liquid Crystals." *Wall Street Journal*, 29 May 1968, 12.

"R.C.A. Elects J. L. Burns as President." *Baltimore Sun*, 16 Jan. 1957, 23.

"RCA Elects John Burns, Business Consultant, President and Director." *Wall Street Journal*, 16 Jan. 1957, 8.

"RCA First Period Sales Set Record: Net Trailed '60." *Wall Street Journal*, 3 May 1961, 11.

RCA Laboratories. *1942–1967: Twenty-Five Years at RCA Laboratories*. Philadelphia: Smith-Edwards, 1967.

"RCA Sales Again Top Billion-Dollar Mark." *Radio Age* 16, no 1 (Jan. 1957): 3–5.

"RCA Says Earnings Rising in 2nd Period: Full-Year High Possible." *Wall Street Journal*, 2 May 1973, 23.

"RCA Says New Product Development Might Make '73 a Record Profit Year." *Los Angeles Times*, 2 May 1973, F9.

"RCA Says There Is 'Good Chance' '61 Sales and Net Will Top '60's." *Wall Street Journal*, 2 June 1961, 6.

"R.C.A. Shows TVs and Radios of '70s." *New York Times*, 10 May 1961, 65, 73.

"RCA's New President." *Radio Age* 16, no. 2 (Apr. 1957): 8.

"RCA to Move Headquarters of 2 Units to Indianapolis." *Wall Street Journal*, 9 Aug. 1960, 4.

"RCA to Open Up 10,000 Radio, TV Patents as It Settles Antitrust Suit." *Wall Street Journal*, 29 Oct. 1958, 3.

"RCA Unveils Models of Electronic Devices to Honor Sarnoff." *Wall Street Journal*, 2 Oct. 1956, 3.

Riordan, Michael, and Lillian Hoddeson. *Crystal Fire: The Birth of the Information Age*. New York: W. W. Norton, 1997.

Rocke, Alan J. *Image and Reality: Kekulé, Kopp, and the Scientific Imagination*. Chicago: University of Chicago Press, 2010.

Rosenbloom, Richard S., and William J. Spencer, eds. *Engines of Innovation: U.S. Industrial Research at the End of an Era*. Boston: Harvard Business School Press, 1996.

Ross, Bill D. "A Bank Loan for 2d-Graders." *New York Times*, 12 May 1974, 79.

Sack, E. A. "ELF: A New Electroluminescent Display." *Proceedings of the IRE* 46, no. 10 (Oct. 1958): 1694–99.

Sarnoff, David. "Electronic Revolution, Present and Future." *New York Times Magazine*, 30 Sept. 1956, 14, 36, 38, 42.

———. "The Fabulous Future." In *The Fabulous Future: America in 1980*, edited by the editors of *Fortune*, 12–31. New York: E. P. Dutton, 1956.

"Sarnoff Challenges Scientists of RCA to Make Three Important Inventions." *Radio Age* 11, no. 1 (Oct. 1951): 5–7, 31.

"Sarnoff Honored: 45 Years in Radio." *New York Times*, 28 Sept. 1951, 32.

Schadt, M., and W. Helfrich. "Voltage-Dependent Optical Activity of a Twisted Nematic Liquid Crystal." *Applied Physics Letters* 18, no. 4 (15 Feb. 1971): 127–28.

Schadt, Martin. "Milestone in the History of Field-Effect Liquid Crystal Displays and Materials." *Japanese Journal of Applied Physics* 48 (2009), doi:org/10.1143/JJAP.48.03B001.

Schindler, H. C. "Liquid-Crystal Dynamic Scattering for Display Devices." *RCA Engineer* 17, no. 6 (Apr.–May 1972): 35–39.

Schmedel, Scott R. "RCA's Plan to Buy Coronet Industries Clears Holders: Minority Objects Loudly." *Wall Street Journal*, 22 Feb. 1971, 6.

"Science Notes: New Laser." *New York Times*, 21 Oct. 1962, E7.

"The Scientist as Inventor and Manager: An Interview with Dr. J. A. Rajchman." *RCA Engineer* 19, no. 2 (Aug.–Sept. 1973): 46–52.

Seeley, P. E. "Electro-Optical Engineering at Burlington." *RCA Engineer* 9, no. 6 (Apr.–May 1964): 63–67.

Shapin, Steven. *The Scientific Life: A Moral History of a Late Modern Vocation*. Chicago: University of Chicago Press, 2008.

Shapiro, Joshua. "George H. Heilmeier." *IEEE Spectrum* 31, no. 6 (June 1994): 56–59.

Shore, Bruce. "Lasers: New Power from Light." *Electronic Age* 21, no. 3 (Summer 1962): 8–11.

———. *The New Electronics*. New York: McGraw Hill, 1970.

———. "'Op' Electronics." *Electronic Age* 27, no. 4 (Autumn 1968): 12–15.

Sluckin, Timothy J., David A. Dunmur, and Horst Stegemeyer, ed. *Crystals That Flow: Classic Papers from the History of Liquid Crystals*. New York: Taylor and Francis, 2004.

Smith, Gene. "Sarnoff's Departure Is Called Result of RCA 'Palace Revolt.'" *New York Times*, 7 Nov. 1975, 49, 53.

Smith, Hedrick. *Rethinking America*. New York: Random House, 1995.

Smith, William D. "RCA Introduces Four New Computers." *New York Times*, 16 Sept. 1970, 63, 69.

Snyder, Samuel S. "Computer Advances Pioneered by Cryptologic Organizations." *Annals of the History of Computing* 2, no. 1 (Jan. 1980): 60–70.

Snyderman, Nat. "Need Improvements: Optel." *Electronic News*, 19 Nov. 1973, 49.

Sobel, Robert. *RCA*. New York: Stein and Day, 1986.

Stashower, Daniel. *The Boy Genius and the Mogul: The Untold Story of Television*. New York: Broadway Books, 2002.

Steele, M. C. "Laboratories RCA, Inc. . . . RCA Research in Japan." *RCA Engineer* 9, no. 1 (June–July 1963): 8–9.

Stephens, Carlene, and Maggie Dennis. "Engineering Time: Inventing the Electronic Wristwatch." *British Journal for the History of Science* 33, no. 4 (Dec. 2000): 477–97.

Stetson, Damon. "'Brain' Accounts for Army's Gear." *New York Times*, 8 Mar. 1957, 27.

Stevens, William K. "Display Devices Crystallize for R.C.A." *New York Times*, 29 May 1968, 47, 54.

Stokes, Donald E. *Pasteur's Quadrant: Basic Science and Technological Innovation*. Washington, DC: Brookings Institution Press, 1997.

Sussman, Alan. "Electrooptic Liquid Crystal Devices: Principles and Applications." *IEEE Transactions on Parts, Hybrids, and Packaging* PHP-8, no. 4 (Dec. 1972): 24–37.

"Swiss Watchmakers Buy American." *Business Week*, 18 Mar. 1972, 21.

Tannas, L. E., Jr. "Electroluminescent Displays." In *Flat-Panel Displays and CRTs*, edited by Lawrence E. Tannas Jr., 237–88. New York: Van Nostrand Reinhold, 1985.

———, ed. *Flat-Panel Displays and CRTs*. New York: Van Nostrand Reinhold, 1985.

———. "Flat-Panel Displays Displace Large, Heavy, Power-Hungry CRTs." *IEEE Spectrum* 26, no. 9 (Sept. 1989): 34–35.

Taylor, George W. "The Design and Operating Characteristics of a 1200-Element Ferroelectric Electroluminescent Display." *IEEE Transactions on Electron Devices* ED-16, no. 6 (June 1969): 565–75.

Taylor, Lynn. "R.C.A. Extinguishes Trademark Flash." *Chicago Tribune*, 19 Jan. 1968, C8.

"Television This Week." *New York Times*, 12 Dec. 1971, D25.

Terry, E. C. "Optel Challenging CRTs with Liquid Crystal Entry." *Electronic News*, 16 Nov. 1970, 62.

Thackray, Arnold, David C. Brock, and Rachel Jones. *Moore's Law: The Life of Gordon Moore, Silicon Valley's Quiet Revolutionary*. New York: Basic Books, 2015.

Thackray, Arnold, and Minor Myers Jr. *Arnold O. Beckman: One Hundred Years of Excellence*. Philadelphia: Chemical Heritage Foundation, 2000.

"The Three C's . . . Electronics' Newest Surge." *Electronic Age* 19, no. 2 (Spring 1960): 13.

"3 Leave Optel: Kiss May Follow." *Electronic News*, 22 Mar. 1976, 34.

Trueb, Lucien F., Günther Ramm, and Peter Wenzig. *Electrifying the Wristwatch*. Atglen, PA: Schiffer, 2013.

Tults, J., and D. L. Matthies. "Facsimile Printer Using a Liquid Crystal Array." *RCA Engineer* 17, no. 6 (Apr.–May 1972): 40–43.

Turkle, Sherry. "'Spinning' Technology: What We Are Not Thinking About When

We Are Thinking about Computers." In *Technological Visions: The Hopes and Fears That Shape New Technologies*, edited by Marita Sturken et al., 19–33. Philadelphia: Temple University Press, 2004.

"TV Firm Plans Suit to Bar Color Move." *New York Times*, 12 Oct. 1950, 50.

"The $20 Digital Watch Arrives a Year Early." *Business Week*, 26 Jan. 1976, 27–28.

Van Raalte, John A. "Reflective Liquid Crystal Television Display." *Proceedings of the IEEE* 56, no. 12 (Dec. 1968): 2146–49.

"Venez à New York le 30 juin! OPTEL dépasse les borne à moins que . . ." *FAN-L'Express*, 22 June 1976, 17.

Warner, J. C. "A History of Radio Corporation of America: The Years to 1938." *RCA Engineer* 3, no. 1 (June–July 1957): 3–9.

"Watch Industry Shifting to Electronic Movements." *Christian Science Monitor*, 21 Aug. 1973, 6.

Weber, Larry F. "History of the Plasma Display Panel." *IEEE Transactions on Plasma Science* 34, no. 2 (Apr. 2006): 268–78.

———. "Plasma Displays." In *Flat-Panel Displays and CRTs*, edited by Lawrence E. Tannas Jr., 332–414. New York: Van Nostrand Reinhold, 1985.

Weimer, P. K. "The TFT . . . A New Thin-Film Transistor," *RCA Engineer* 7, no. 6 (Apr.–May 1962): 20–22.

Werner, Ken. "The Flowering of Flat Displays." *IEEE Spectrum* 34, no. 5 (May 1997): 40–49.

Williams, Richard. "Domains in Liquid Crystals." *Journal of Chemical Physics* 39, no. 2 (15 July 1963): 384–88.

———. "Effect of a High Electric Field on the Absorption of Light by PbI_2 and HgI_2." *Physical Review* 126, no. 2 (15 Apr. 1962): 442–46.

———. "Electric Field Induced Light Absorption in CdS." *Physical Review* 117, no. 6 (15 Mar. 1960): 1487–90.

———. "High Electric Fields in Sodium Chloride." *Journal of Physics and Chemistry of Solids* 25, no. 8 (Aug. 1964): 853–58.

———. "Liquid Crystals in an Electric Field." *Nature* (20 July 1963): 273–74.

Williams, Richard, and Richard H. Bube. "Photoemission in the Photovoltaic Effect in Cadmium Sulfide Crystals." *Journal of Applied Physics* 31, no. 6 (June 1960): 968–78.

Williams, Richard, and George Heilmeier. "Possible Ferroelectric Effects in Liquid Crystals and Related Liquids." *Journal of Chemical Physics* 44, no. 2 (15 Jan. 1966): 638–43.

Wisnioski, Matthew. *Engineers for Change: Competing Visions of Technology in 1960s America*. Cambridge, MA: MIT Press, 2012.

Woolley, Scott. *The Network: The Battle for the Airwaves and the Birth of the Communications Age*. New York: HarperCollins, 2016.

Zaki, Tarek. *Short-Channel Organic Thin-Film Transistors*. Cham, Switzerland: Springer, 2015.

"Zenith and RCA Settle $61 Million Suit Out of Court." *Wall Street Journal*, 10 Sept. 1957, 12.

Zipser, Alfred R. "Move by Zenith Hailed by R.C.A." *New York Times*, 24 Feb. 1961, 37, 45.

———. "Personality: Pilot for R.C.A. in Space Field." *New York Times*, 17 May 1959, F3.

———. "R.C.A. Sales of Data Systems Running 200% above '59 Level." *New York Times*, 25 Sept. 1960, F1, F24.

———. "R.C.A. Stepping Up Computer Making." *New York Times*, 14 Apr. 1957, F1, F9.

INDEX

Numbers in italics refer to illustrations.